Rainwater Harvesting for Agriculture in the Dry Areas

Rainwater Harvesting for Agriculture in the Dry Areas

Editor

Mohd. Naushad Ansari

scitus
academics

Rainwater Harvesting for Agriculture in the Dry Areas

Edited by **Mohd. Naushad Ansari**

ISBN: 978-1-68117-016-9

Library of Congress Control Number: 2015931390

SCITUS Academics LLC,
www.scitusacademics.com
Box No. 4766, 616 Corporate Way,
Suite 2, Valley Cottage,
NY 10989

Printed in United States of America on Acid Free Paper ∞

Contents

Preface vii

Chapter 1 **Macro Rain Water Harvesting Network to Estimate Annual Runoff at Koysinjaq (Koya) District, Kurdistan Region of Iraq** 1
Saleh Zakaria, Nadhir Al-Ansari, Yaseen T. Mustafa, M. D. J. Alshibli, Sven Knutsson

Chapter 2 **Solar Disinfection of Pseudomonas aeruginosa in Harvested Rainwater: A Step towards Potability of Rainwater** 25
Muhammad T. Amin, Mohsin Nawaz, Muhammad N. Amin, Mooyoung Han

Chapter 3 **Microbial Quality and Phylogenetic Diversity of Fresh Rainwater and Tropical Freshwater Reservoir** 51
Rajni Kaushik, Rajasekhar Balasubramanian, Hugh Dunstan

Chapter 4 **Dryland Soil Hydrological Processes and Their Impacts on the Nitrogen Balance in a Soil-Maize System of a Freeze-Thawing Agricultural Area** 79
Wei Ouyang, Siyang Chen, Guanqing Cai, Fanghua Hao

Chapter 5 **Two Challenges for U.S. Irrigation Due to Climate Change: Increasing Irrigated Area in Wet States and Increasing Irrigation Rates in Dry States** 109
Robert I. McDonald, Evan H. Girvetz

Chapter 6 **Aquatic Ecosystem Response to Timber Harvesting for the Purpose of Restoring Aspen** 137
Bobette E. Jones, Monika Krupa, Kenneth W. Tate

Chapter 7 **Evidence of a High Density Population of Harvested Leopards in a Montane Environment** 197
Julia N. Chase Grey, Vivien T. Kent, Russell A. Hill

Chapter 8 **Climate Change and Challenges of Water and Food Security** 225
Anil Kumar Misra

Chapter 9 **Burnt Sugarcane Harvesting – Cardiovascular Effects on a Group of Healthy Workers, Brazil** **257**

Cristiane Maria Galva˜o Barbosa, Ma´ rio Terra-Filho, Andre´ Luis Pereira de Albuquerque, Dante Di Giorgi, Cesar Grupi, Carlos Eduardo Negra˜o, Maria Urbana Pinto Branda˜o Rondon, Daniel Godoy Martinez, Ta^ nia Marcourakis, Fabiana Almeida dos Santos, Alfe´ sio Luı´s Ferreira Braga, Dirce Maria Trevisan Zanetta, Ubiratan de Paula Santos

Citations **285**

Index **289**

Preface

Only a tiny portion of the rainfall in semi-arid and arid areas falls into deeper soil or maybe rock layers to recharge an aquifer. Another small percentage is used for vegetation transpiration or possibly farm crops. Most of the precipitation evaporates from the often bare soil or depressions of the surface. More irrigation is needed to feed the growing population in the dry regions of the world, but the amount of irrigation water is very small. The "classic" sources of irrigation water are often overused and unused sources of irrigation water must therefore be sought. The use of the evaporative portion of precipitation to be used for agricultural purposes before it is released into the atmosphere is necessary to increase agricultural production in dry areas. Water harvesting is a cost-effective, simple and environmentally friendly way to recover a large part of the lost water. There is a need to use and capture the scarce water resources more effectively in low rainfall regions.

Optimizing rainfall management by harvesting water in integrated and sustainable production systems can help to improve the livelihood of small-scale farmers by improving rainfed farm production. This book is a compilation of the state-of-the-art information on water harvesting techniques, which have been tested and found to be useful somewhere and can be used in other conditions. The book also attempts to show some successful cases of the use of water harvesting techniques in low rainfall areas, which have improved the overall productivity of smallholder farms and thus improved the livelihood of farmers.

Chapter 1

Macro Rain Water Harvesting Network to Estimate Annual Runoff at Koysinjaq (Koya) District, Kurdistan Region of Iraq

Saleh Zakaria[1,2], Nadhir Al-Ansari[2*], Yaseen T. Mustafa[3], M. D. J. Alshibli[4], Sven Knutsson[2]

[1]Department of Water Resources Engineering, University of Mosul, Mosul, Iraq

[2]Department of Civil, Environmental & Natural Resources Engineering, Lulea University of Technology, Lulea, Sweden

[3]University of Zakho, Duhok, Iraq

[4]Ministry of Water Resources, Baghdad, Iraq

ABSTRACT

Macro rainwater harvesting techniques (Macro RWH) are getting more popular to overcome the problem of water scarcity in arid and semi-arid areas. Iraq is experiencing serious water shortage problem now despite of the presence of Tigris and Euphrates Rivers. RWH can help to overcome this problem. In this research, RWH was applied in Koya City in its districts, North West Iraq. Twenty-two basins were identified as the catchment area for the application of RWH technique. Watershed modeling system (WMS), based on Soil Conservation Service-curve number (SCS-CN) method, was applied to calculate direct runoff from individual daily rain storm using average annual rainfall records of the area. Two consecutive adjustments for the curve number were considered. The first was for the antecedent moisture condition (AMC) and the second was for the slope. These adjustments increased the total resultant harvested runoff up to 79.402×10^6 m^3. The average percentage of increase of harvested runoff volume reached 9.28%. This implies that water allocation is of the order of 2000 cubic meter per capita per year. This quantity of water will definitely help to develop the area.

INTRODUCTION

The Middle East and North Africa (MENA) region is characterized by its arid to semiarid climate where the average annual rainfall does not exceed 166 mm [1]. Iraq is part of the MENA region and was not facing any water shortages till the 1970s. After that, the dams built on the upper parts of the Tigris and Euphrates Rivers in Syria and Turkey plus the effect of global warming had changed the situation [2, 3]. In evaluating Iraqi water resources issue, the future predictions suggest more shortages [4, 5]. It is expected that Tigris and Euphrates Rivers will be completely dry by 2040 [6].

Furthermore rainfall is not sufficient to support economic crop yield during rainy seasons without irrigation. The average annual rainfall in Iraq is ranging from 154 to 216 mm/year [7, 8]. It should be mentioned that the rainfall widely varies from north to south and from west to east of Iraq, where it reached more than 1000 mm within the mountains at the north, 150 mm within the western desert to about 200 mm at the eastern part of the country. In many regions having limited water resources, including surface or sub-surface water, the available water is no longer sufficient to cover the ever increasing water demand [9]. For this reason, farmers are using groundwater in irrigation to cover the shortages due to low rainfall. As a consequence, excessive pumping of ground water was practiced, which led to falling of water tables in different parts of the Middle East [10, 11]. Thus water scarcity will be one of the major challenges facing the world during this century [12] and the Middle East in particular [13,14].

The limitation of water sources, rising water demand in addition to mismanagement water resources, in Developing World, is contributing to the water scarcity problem [12].

Water resource management is becoming one of the most important economic and social issues in this century [15]. Therefore, the situation requires a new technique and method for conservation and judicious water [12].

Some countries of the MENA region had provided alternative non-traditional water sources, such as rainwater harvesting (RWH), to overcome the water scarcity problem [16-18].

In this research, RWH technique is to be used to overcome the water shortage problem in Koya area, northeast of Iraq. RWH has different definitions [19]. Boers and Ben-Asher, (1982) [20] gave a more specific definition of RWH with specific details, where they defined RWH as "a method for inducing, collecting, storing, and conserving local surface runoff for agriculture in arid and semi-arid regions". They explained that the RWH includes several processes which start dealing with the catchment area to guarantee the stream runoff flow then to direct it by the natural drainages that distributed on the catchment area to the target storage location (a surface reservoir or a soil profile). Furthermore, they specified the aim of this process for agriculture purposes. Finkel and Finkel, (1986) [21] defined RWH as "the collection of runoff and its use for the irrigation of crops, pastures and trees, and for livestock consumption". Siegert, (1994) [22] defined RWH as "the collection of runoff for its productive use".

Prinz, (2000) [12] summarized six different forms of RWH according to the location, function and size of catchment area as follows: 1) Roof Top RWH, 2) RWH for Animal Consumption, 3) Inter-Row RWH, 4)

Micro-catchment RWH, 5) Medium-sized Catchment RWH and 6) Large Catchment RWH (Macro-catchment).

For the annual rainfall between 100 and 700 mm, water harvesting might provide new source of water which is not readily available or too costly [23].

The productivity of the rainwater can be significantly improved by applying a specific technique such as Macro RWH, based on availability of a surface reservoir. By this technique, the excess rainwater (runoff) is stored in small reservoirs of small dams with different sizes to be supplied later when required [24-27].

RWH can only increase the availability of rainwater to the user but not its amount, certainly by concentrating the excess rainwater (runoff) in a limited area which increases the potential risk of erosion, so suitable measures must be taken to prevent soil erosion [28].

RWH systems had proven to be an effective technique in different regions to achieve new water source that can be used for several purposes, furthermore, in comparison with pumping water, water harvesting saves energy and maintenance costs [23].

Macro catchment RWH systems gave good results at different parts of the world and led to an increase in crop production [29-31]. Furthermore, studies by some researchers (e.g. Bruins *et al.*, 1986; Fox and Rockstrom, 2003; Hatibu *et al.*, 2003; Motsi *et al.*, 2004; Barron and Okwach, 2005; Liu *et al.*, 2005 [32-37]) were conducted in different parts of Africa which indicated that rainwater harvesting is working to reduce the risk of drought and increase agricultural production. For the above studies, RWH generates a new source of water where water is not readily available [38]. Hatibu and Mahoo, (1999) [39], indicate that Macro RWH is a system that involves the collection of runoff from large areas that are ranging from 0.1 ha to thousands of hectares with slopes ranging from 5% to 50%. This system is used in Tanzania with storage of water outside the cropped basin for later use.

Most of the techniques of water harvesting systems focus on capturing more water [40], for Macro RWH, however, it is the capture rainfall that falls outside the farmland [41]. Effective management of RWH becomes more interesting phase of water resources management strategies in most countries that are suffering from the problem of water scarcity.

RWH is the use of lost runoff water and it proved to be one of the most effective methods to overcome water shortages in arid and semi-arid regions [19-22,27,28]. In addition, Macro RWH significantly improves the productivity and it increases the availability of rainwater to the user and it was proved that this system gave very good results [24,29,30]. In view of the above, this method was applied in Koya and its districts.

This research is treating RWH for the whole area that is composed of twenty-two selected sites. It is anticipated to establish a network of Macro RWH distributed around Koya City, Kurdistan region of Iraq in order to estimate the annual amount of runoff that could be harvested and used. Furthermore, this type of technique and the modified curve numbers had been used for the first time in this area.

METHODOLOGY

Dams are to be built to harvest the excess rainfall (runoff) for a given area. To achieve this goal then it is necessary to identify the sites of the dams. This can be done using Digital Elevation Model (DEM) of the study area with Global Mapper model. Certainly the dam site will be located on runoff stream (main trajectory) considering minimum dam's cross section to minimize the constructions cost. Watershed modeling system (WMS) was used with Konya's DEM. The model was applied using the information obtained from land use map, soil type and rainfall data, for all individual selected basins. The runoff volumes were estimated based on Soil Conservation Service-curve number (SCS-CN) method. The main steps that should be follow can be summarized as follow: Identification each of the drainage boundaries of the selected basins within the study area, the hydrologic soil group classification to determine the runoff curve number for a given soil kind using the tables of SCS, 1972. Then land use map is usually used to identify the curve number (CN) values for each selected basin. The time of concentration for selected basins is estimated, and the daily rainfall depth is to be determined considering single rainfall storm on the study area. Then the suitable storm type (I, 24-hour Storm) should be chosen. The hydrograph time increments of six minutes or less are to be set, and the volumes of runoff for the selected basins are to be calculated.

Furthermore, the SCS curve number method based on the relationships between precipitation and runoff expressed as:

$$Q = \frac{(p - 0.2S)^2}{P + 0.8S} \text{ if } p > 0.2S \tag{1}$$

$$Q = 0 \text{ if } p \le 0.2S \tag{2}$$

$$Q = \left(\frac{1000}{CN} - 10 \right) * 25.4 \tag{3}$$

where:

Q = the direct runoff or rainfall excess (mm).

P = the storm rainfall (mm).

S= the maximum potential soil water retention (mm), and *CN* = the curve number (dimensionless).

With SCS-CN method, the soil was classified into four hydrological soil groups A, B, C and D considering the basin wetness index i.e. the antecedent moisture condition (AMC) which had been classified into three classes AMC I, AMC II and AMC III, representing dry, average and wet conditions. In order to specify each class, the antecedent rainfall amount of five-day and season category (dormant and growing seasons) were considered.

For the Soil Conservation Service, 1972 (SCS-CN) method, the tabulated curve number is equal to CNII, for average (normal) conditions, and modified for dry and wet conditions, as explained by Chow [42] through the following equations:

$$CN_I = \frac{4.2 * CN_{II}}{10 - 0.058 * CN_{II}} \tag{4}$$

$$CN_{III} = \frac{23 * CN_{II}}{10 + 0.13 * CN_{II}} \quad (5)$$

In which: CN_I = Curve number for dry condition.

CN_{III} = Curve number for wet condition.

Williams [43] developed an equation to adjust the curve number to a different slope [44,45]:

$$[CN_{II}]_{SLP} = \frac{(CN_{III} - CN_{II})}{3} * [1 - 2 * \exp(-13.86 * SLp)] + CN_{II} \quad (6)$$

where:

$[CN_{II}]_{SLP}$ = the curve number for average condition adjusted for the slope.

SLP = the average fraction slope of the basin.

Study Area

Koysinjaq (Koya) is one of the most important districts of Erbil Governorate at Kurdistan region of Iraq (Figure 1), which is witnessing a state of rapid growth and development. According to the Iraqi statistics of 1987 the population of Koysanjaq is about 39,484 people.

Koya districts have a very important geographical location where it connects three provinces of Iraq which are Sulaimaniyah, Kirkuk and Erbil. Koya district consists mainly of five parts which are TaqTaq, Ashti, Shoresh, Sktan and Sekrkan. The district is bordered from the east and south by lesser Zab River, and from the northeast by Hebat Sultan mountain, from the west by Bawage Mountain. The mountainous area is located north Koysinjaq, while at the south and southwest, a fertility plain extends to the border of Erbil with Kirkuk city, which represents the historical alluvial plain of the Tigris River. Rainwater is the main source for agricultural processes in the area in addition to the ground water.

The soil texture in the mountainous regions is sandy clay, loam silt or loam clay sand, with an average depth of 130 cm. While the soils texture of the plain regions consists of loam clay sand, loam silt and silt clay, with an average depth of 140 cm [46]. Buringh (1960) [47] described the soil of study area (Figure 2) and refer that soil color varies between light yellow to dark brown at north and between brown and dark brown at the plain regions as a shallow phase over Bakhtiary gravel.

Koya Rainfall

Rainfall records at Koya station of the period 2002-2003 to 2010-2011 were used in this research. Two seasons (2007-2008 and 2008-2009) where neglected due to missing data. These records show that the rainy season extends from November to May. The annual rainfall varies from one season to the other. The total rainfall reached minimum value of 433.9 mm during the season 2005-2006, while it reached maximum value of 989.2 mm during the season 2003-2004 (Figure 3).

The average rainfall depth for the study period reached 650.2 mm which is very close to that of the season 2009- 2010.

Figure 4, shows that the fourteen rain storms with appropriate antecedent moisture as recommended by SCS-CN method for the season 2009-2010.

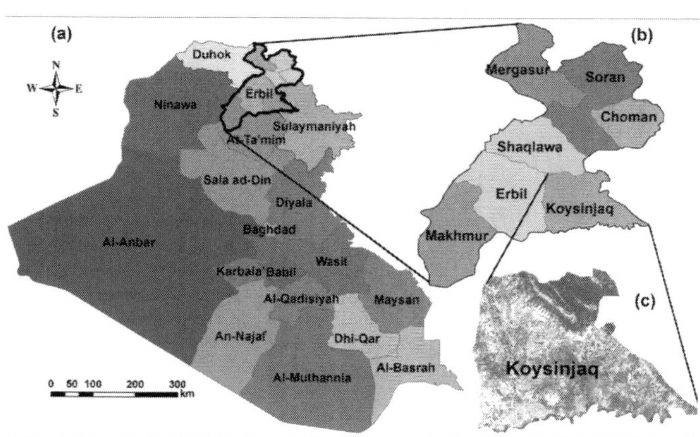

Figure 1: Map of Iraq (a); Erbil governorate (b); Location of study area (Koya) (c).

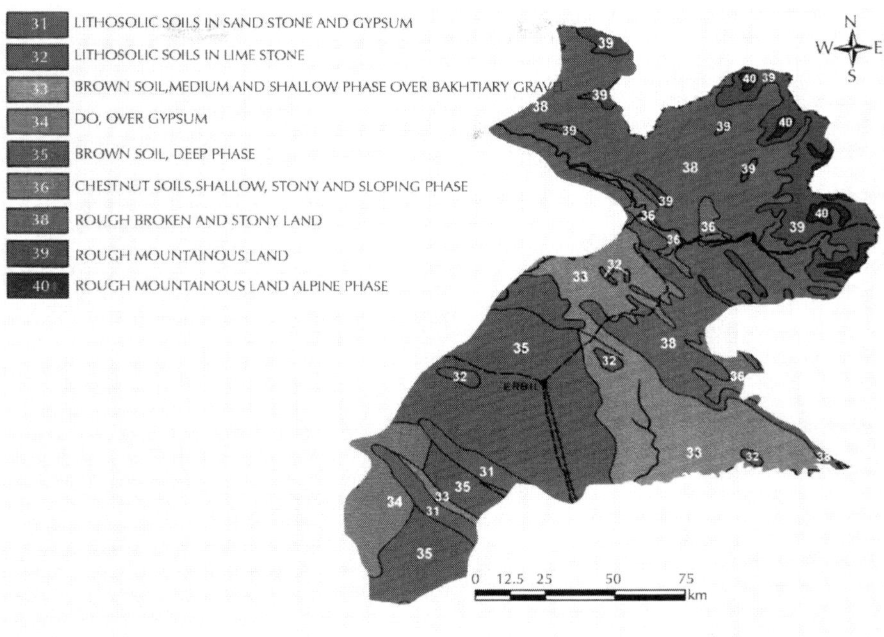

Figure 2: Soil map of Erbil Government as described by Buringh 1960, source [48].

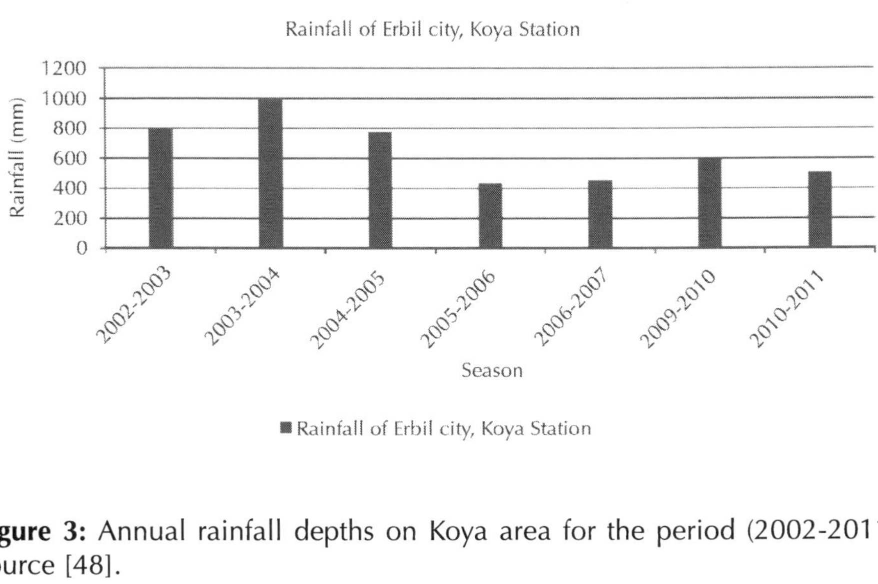

Figure 3: Annual rainfall depths on Koya area for the period (2002-2011), source [48].

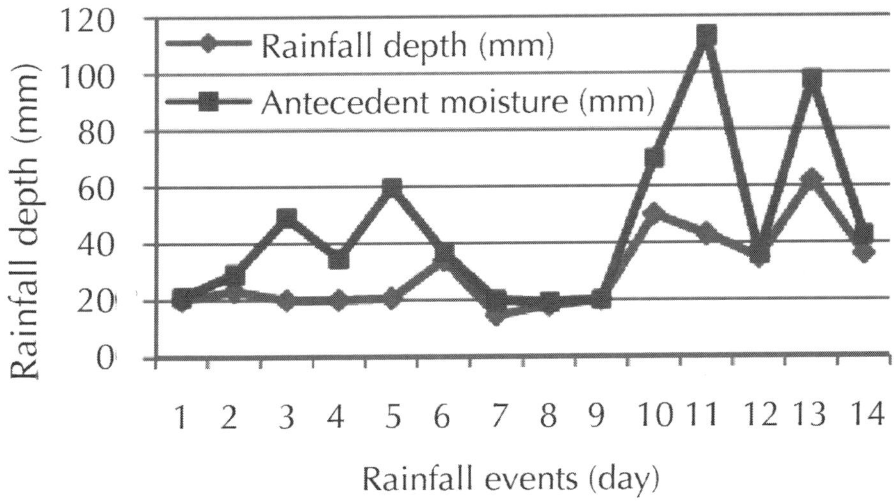

Figure 4: Rainfall depth with corresponding antecedent moisture for the season (2009-2010).

The rainfall season 2009-2010 starts at the first of November. The season includes fifty-six of daily rain storms that were distributed along 210. Fourteen of these rainfall storms had produced runoff events. Of these, four storms were under average conditions and ten under wet conditions. These conditions were specified according to the antecedent moisture classes (AMC) for the SCS-CN method.

The study has been focused on the above fourteen rain storms that each has exceeded 12.5 mm in depth and produced runoff. Sequences were given to the rain storms according to the time of occurrence.

Land Use/Land Cover

Figure 5 shows that land use land cover (LULC) map for Koya district with twenty-two selected basins for rainwater harvesting. LULC map was classified into five classes (type) in the study area. They were building up, vegetation, bare soil, rock, and water. The basins were numbered starting from the far north of Koya city in an anticlockwise direction.

Curve Number

The weighted average CN values for twenty-two selected basins at Koya District were estimated depending on area of specific land use land cover as a percent of total basin area and calibrated based on antecedent moisture condition (AMC) for dry, average, and wet conditions depending on the total antecedent rainfall depth of five days as formulated by Soil Conservation Service-curve number (SCS-CN) method. The properties of the selected basins at Koya District were estimated (Table 1).

Then CN values were adjusted for slope using Williams [43] formula for each basin (Table 2).

RESULTS AND DISCUSSION

Data elevation modeling (DEM) of Koya districts of Kurdistan region of Iraq, was used to identify the suitable sites of dams in order to harvested the exceed rainwater (runoff) from the catchments area. Twenty-two basins were selected for rainwater harvesting.

It should be noted that any basin that was not representing a continuous hydrological unit was excluded. In other word, the runoff should be continuously drained within its catchment area to the outlet where the harvested dams are located.

The harvested runoff from the individual selected basins can be stored at the outlet of each basin to conform individual reservoirs of different capacities using harvested dams. The harvested runoff volumes were estimated using WMS.

To simplify the analysis of the harvested runoff, the twenty-two selected basins at Koya District were divided into four groups of basins according to the geographic- location as follows: At the north, group number one includes the basins 1B, 2B, 3B, 4B, and 5B. Southwest of Koya, group number two includes 6B, 7B, 8B, 9B, 10B, 11B, 12B and 13B. At the middle south of Koya district, group number three includes 14B, 15B, 16B, and 17B. At the southeast, group number four includes 18B, 19B, 20B, 21B, and 22B.

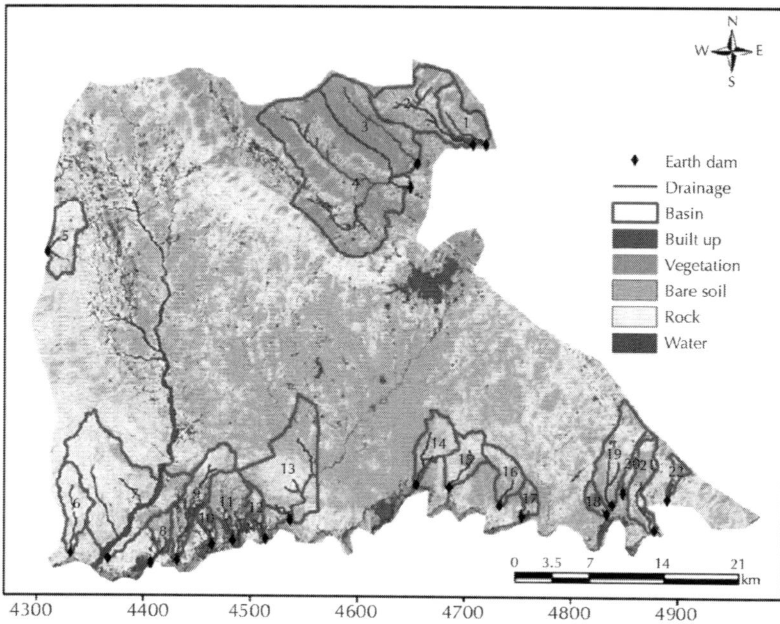

Figure 5: LULC map for Koya districts with twenty-two selected basins for rainwater harvesting, source [48].

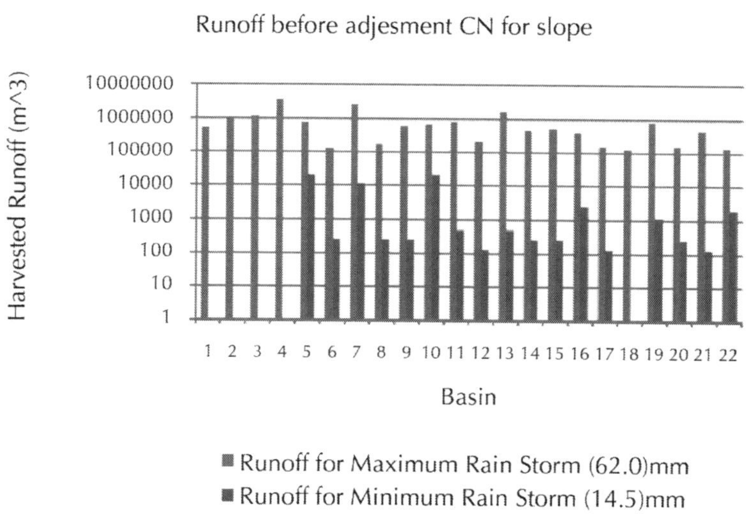

Figure 6: Harvested runoff volumes for maximum and minimum rain storms, before and after adjustment Curve Number for the slope for the 22 basins.

Table 1: Properties of twenty-two selected basins at Koya districts

Group	Basins	Area (Km^2)	Slope (Sin)	Length (m)	Elevation (m)	Basin area ratio
	1	30.95	0.15	29095.29	684.3	6.26
	2	15.51	0.14	6771.437	740.7	3.14
1	3	37.68	0.28	26771.5	845.8	7.63
	4	111.63	0.19	15822.17	913.8	22.59
	5	17.15	0.08	5552.237	740.7	3.47
	6	15.57	0.08	8464.906	554.1	3.15
	7	72.99	0.08	13505.08	426.4	14.77
	8	6.22	0.06	5245.913	361.5	1.26
	9	22.14	0.08	11863.12	436.8	4.48
2						
	10	3.47	0.06	630.3264	367.3	0.70
	11	17.95	0.09	6697.066	407.2	3.63
	12	4.90	0.08	5648.858	413.6	0.99
	13	47.40	0.09	9330.842	407.5	9.59
	14	14.12	0.09	7112.508	518.5	2.86
	15	13.13	0.14	6021.934	584.9	2.66
3						
	16	11.53	0.11	5957.011	531.9	233
	17	4.45	0.08	3552.444	472.4	0.90
	18	3.86	0.09	3534.156	447.4	0.78
	19	22.14	0.12	9910.267	634.3	4.48
4	20	4.43	0.10	6418.783	632.8	0.90
	21	13.55	0.12	8739.835	548.9	2.74
	22	3.34	0.20	4392.473	711.4	0.68

Table 2: Curve number (CN) values before and after adjustment for slope for the twenty-two selected basins at Koya districts

	Before adjustment CN for slope				After adjustment CN for slope	
Basins	Normal	dry	wet	Normal	dry	wet
1	76.0	57.1	87.9	78.97	61.20	89.63
2	75.9	56.9	87.8	78.73	60.85	89.49
3	75.1	55.9	87.4	79.03	61.28	89.66
4	76.0	57.1	87.9	79.40	61.81	89.86
5	86.5	73.0	93.7	87.32	74.30	94.06
6	87.0	73.7	93.9	87.78	75.11	94.29
7	80.7	63.6	90.6	81.82	65.40	91.19
8	78.4	60.3	89.3	78.87	61.05	89.57
9	78.5	60.5	89.4	79.74	62.30	90.05
10	79.6	62.2	90.0	80.05	62.76	90.22
11	77.1	58.6	88.6	78.73	60.86	89.49
12	77.1	58.6	88.6	78.40	60.39	89.30
13	78.0	59.8	89.1	79.57	62.07	89.96
14	77.7	59.4	88.9	79.29	61.65	89.80
15	81.1	64.3	90.8	83.40	67.85	92.04
16	78.4	60.4	89.3	80.45	63.35	90.44
17	77.5	59.1	88.8	78.78	60.93	89.52
18	75.4	56.3	87.6	77.13	58.62	88.58
19	79.1	61.3	89.7	81.29	64.61	90.91
20	79.1	61.3	89.7	80.87	63.96	90.67
21	77.9	59.6	89.0	80.20	62.98	90.31
22	84.0	68.8	92.4	86.45	72.82	93.62

The area of the twenty-two selected basins ranged 3.34 - 111.63 km^2 and the total area of the selected basins is 494.11 km^2. Basins slope ranged between 6% - 28%, their length ranged between 0.63 - 29.09 km, and their elevation ranged bet we en 361.5 - 913.8 m (Table 1).

In spite of the fact, that, most of the selected basins are small in their areas, but their runoff is of relatively good quantity. However, the area is not the only decisive factor to control the quantity of runoff, although it is one of the important factor to maximize the volume of runoff in the basin, but still other factors like CN values (which represent the hydraulic conditions of the selected land) and the slope are more sensitive to reflect their strong impact on the composition of runoff. It is very difficult to separate the effect of the variables involved (e.g. area, slop and CN) on the produced runoff at a given basin.

However, these factors in addition to the rainfall pattern play an important role together to form the harvested runoff. In fact, rainfall has two effects, the first is its amount, so as far as there is an increase in rainfall depth that will help to increased harvested runoff amount.

Secondly, by its distribution *i.e.* when the span time (between two subsequent rain storms) increases or decreases.

A comparison of runoff volumes for all sloped basins, under same condition, may explain the effect of rainfall depth, and basins' slop. The runoff was always achieved in maximum volumes under maximum rain storm (62.0 mm) and minimum volume under minimum rain storm (14.5 mm) also the runoff increased after adjusted CN for slope (Figure 6).

It should be noted that, the weak rainfall storm (that does not produce runoff) is very important for estimating the CN values. The weak rainfall effects directly the antecedent moisture condition (AMC) and then the corresponding value of CN an d change its value from average to wet condition or vice versa and this is very sensitive for runoff calculations [48].

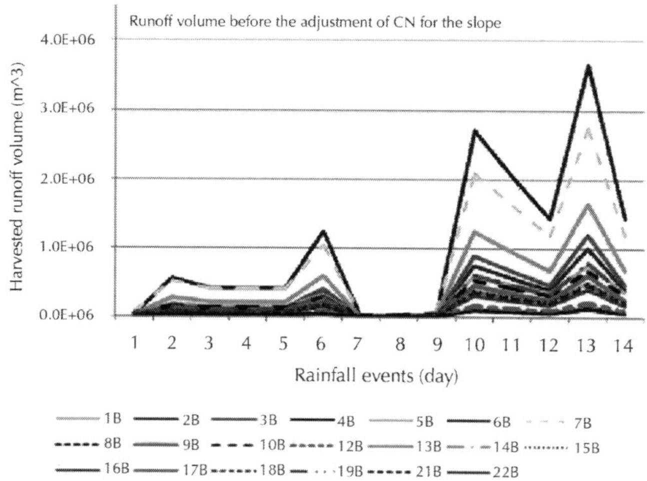

Figure 7: Seasonal harvested runoff through fourteen rain storms events for the selected basins.

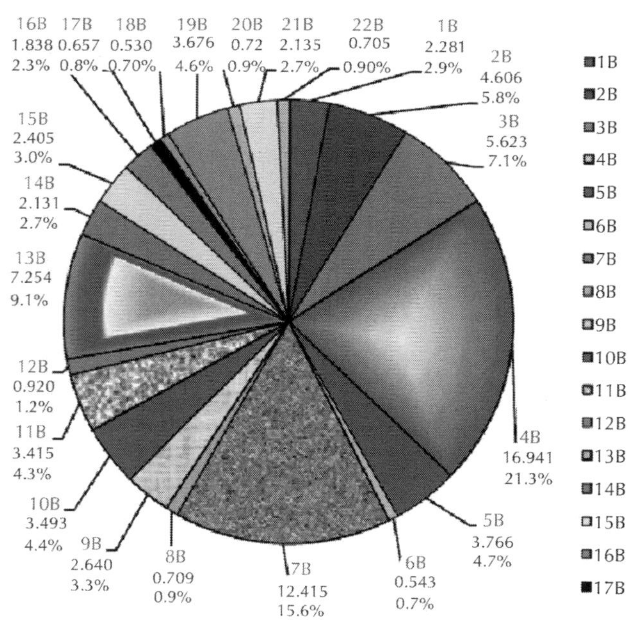

Figure 8: Annual harvested runoff that contributed by each basin as a volume ($\times 10^6$ m^3) with its percentage of total annual harvested runoff at total Koya districts, considering the adjustment CN for the slope.

Figures 7 and 8 show the results of harvested runoff by all groups of the basins. The harvested runoff volume, without adjusted CN for the slope, reached up to 28.618, 29.543, 6.421 and 7.005 million cubic meters for the group 1 to 4 respectively. With adjusted CN for the slope, the harvested runoff reached up to 33.217, 31.388, 7.031, and 7.766 million cubic meters for the group 1 to 4 respectively. The results of total harvested runoff by all 22 basins were 71.586 and 79.402 million cubic meters before and after the adjustment of CN for the slope respectively. Within the selected basins, the maximum harvested runoff achieved at basin 4B were 14.373 and 16.941 million cubic meters and minimum at basin 18B were 0.485 and 0.530 million cubic meters before and after the adjustment of CN for the slope respectively.

Figure 7 shows that the comparison of harvested runoff volumes between two cases (before and after) the adjustment CN for the slope. The comparison shows that the runoff patterns for all fourteen rainstorms that produced runoff.

The results indicate that there was an increase in harvested runoff volume due to the adjustment of CN for the slope. The maximum, minimum, and average increase of harvested runoff volume reached 20.81%, 1.92%, and 9.28% respectively.

Figure 8 shows that the annual harvested runoff contribution of each basin as a volume with its percentage of total harvested runoff volume at total Koya districts considering the adjustment of CN for the slope, where the maximum harvested runoff was achieved at basin 4B (16.941) million cubic meters represented 21.3% of total annual harvested runoff volume, while the minimum harvested runoff was achieved at basin 18B (0.530) mil lion cubic meters represented 0.7 % of total annual harvested runoff volume.

The total quantity of water (79.402 million cubic meters) if harvested will give an annual allocation of about 2000 cubic meter per capita. In addition, hundreds of square kilometers of land can be irrigated using the harvested water.

CONCLUSIONS

Koya City and its districts, at Kurdistan region of Iraq, are rapidly developing under conditions of limited water availability. All future expectations indicate more severe shortages in water resources in Iraq. It is believed that rain water harvesting technique can help a large extent to overcome this situation. The results obtained with water harvesting technique using the average annual rainfall showed that a minimum of 79.402×10^6 cubic meters of water can be harvested annually. This suggests that the allocation per capita per year will be about 2000 cubic meters. This will definitely help to develop the agricultural and industrial activities in the area.

The results indicated that there was an increase in harvested runoff volume due to the adjustment of CN for the slope. The maximum, minimum, and average percentage of increase of harvested runoff volume reached 20.81%, 1.92%, and 9.28% respectively.

The results show that Koya district has the ability to produce good amount of annual volume of runoff that reached, but unfortunately, most of these quantities are lost without any benefit.

ACKNOWLEDGEMENTS

The authors would like to express their sincere thanks to the Ministry of Higher Education and Scientific Research, Baghdad, Iraq; Mosul university, Mosul, Iraq for their support. Deep thanks to Mrs. Semia Ben Ali Saadaoui of the UNESCO-Iraq for her encouragement and support. The research presented has been financially supported by Luleå University of Technology, Sweden and by "Swedish Hydropower Centre—SVC" established by the Swedish Energy Agency, Elforsk and Svenska Kraftnät together with Luleå University of Technology, The Royal Institute of Technology, Chalmers University of Technology and Uppsala University. Their support is highly appreciated.

REFERENCES

1. N. A. Al-Ansari, "Water Resources in the Arab Countries: Problems and Possible Solutions," *UNESCO International Conference (Water: A Looming Crisis)*, Paris, 3-6 June 1998, pp. 367-376.
2. N. A. Al-Ansari and S. Knutsson, "Toward Prudent Management of Water Resources in Iraq," *Journal of Advanced Science and Engineering Research*, Vol. 1, 2011, pp. 53-67.

3. N. A. Al-Ansari, "Management of Water Resources in Iraq: Perspectives and Prognoses," *Engineering*, Vol. 5, No. 8, 2013, pp. 667-684. http://dx.doi.org/10.4236/eng.2013.58080.
4. N. A. Al-Ansari, E. Salameh and I. Al-Omari, "Analysis of Rainfall in the Badia Region, Jordan," Al-Bayt University Research Paper No.1, 1999, 66 p.
5. A. Hamdy, "Water Crisis and Food Security in the Arab World: The Future Challenges," 2013. http://gwpmed.org/files/IWRM-Libya/Atef%20Hamdy%20AWC.pdf
6. United Nations, "Water Resources Management White Paper," United Nations Assistance Mission for Iraq, United Nations Country Team in Iraq, 2010, 20 p. http://iq.one.un.org/documents/100/white%20paper-eng_Small.pdf
7. N. A. Al-Ansari, H. I. Assaid and V. N. Salim, "Water Resources in Iraq," *Journal of the Geological Society*, Vol. 15, 1981, pp. 35-42.
8. FAO, "Water Profile of Iraq," 2008. http://www.eoearth.org/view/article/51cbef2d7896bb431f69ceb3/
9. D. Prinz, "The Role of Water Harvesting in Alleviating Water Scarcity in Arid Areas," Keynote Lecture, *Proceedings of International Conference on Water Resources Management in Arid Regions*, Vol. III, Kuwait, Kuwait Institute for Scientific Research, 23-27 March 2002, pp. 107-122.
10. J. Chenoweth, P. Hadjinicolaou, A. Bruggeman, J. Lelieveld, Z. Levin, M. Lange, E. Xoplaki and M. Hadjikakoul, "Impact of Climate Change on the Water Resources of the Eastern Mediterranean and Middle East Region: Modeled 21st Century Changes and Implications," *Journal of Water Resources Research*, Vol. 47, 2011, Article ID: W06506, pp. 1-18.
11. K. Voss, J. Famiglietti, M. Lo, C. de Linage, M. Rodell and S. Swenson, "Groundwater Depletion in the Middle East from GRACE with Implications for Transboundary Water Management in the Tigris-Euphrates-Western Iran Region," *Water Resources Research*, Vol. 49, No. 2, 2013, pp. 904-914. http://dx.doi.org/10.1002/wrcr.20078
12. D. Prinz and A. K. Singh, "Water Resources in Arid Regions and Their Sustainable Management. Special Issue on Research," *Annals of Arid Zone*, Vol. 39, No. 3, 2000, pp. 1-8.

13. T. Naff, "Conflict and Water Use in the Middle East," In: R. Roger and P. Lydon, Eds., *Water in the Arab Word: Perspectives and Prognoses*, Harvard University, 1993, pp. 253-284.
14. N. A. Al-Ansari, "Applied Surface Hydrology," Al-Bayt University Publication, Mafraq, 2005.
15. K. Shideed, T. Oweis and E. Mohammad, "Enhancing Agricultural Productivity through On-Farm Water-Use Efficiency: An Empirical Case Study of Wheat Production in Iraq," ESCWA/SDPD, New York, 2003, p. 15.
16. K. K. Gupta, J. Deelstra and K. D. Sharma, "Estimation of Water Harvesting Potential for a Semiarid Area Using GIS and Remote Sensing, Remote Sensing and Geo-graphic Information System for Design and Operation of Water Resources System," *Proceedings of Information Symposium*, Rabat, Vol. 242, 1997, pp. 53-62.
17. S. Nasri, A. Jean, C. Christophe and B. Ronny, "Hydrological Processes in Macro Catchment Water Harvesting in the Arid Region of Tunisia: The Traditional System of Tabias," *Journal of Hydrological Sciences*, Vol. 49, No. 2, 2004, pp. 261-272. http://dx.doi.org/10.1623/hysj.49.2.261.34838
18. T. Oweis, A. Hachum and J. Kijne, "Water Harvesting and Supplemental Irrigation for Improved Water Use Efficiency in Dry Areas," International Water Management Institute, Colombo, 1999, 41 p. http://www.iwmi.cgiar.org/Publications/SWIM_Papers/PDFs/Swim07.pdf
19. C. Reij, P. Mulder and L. Begemann, "Water Harvesting for Plant Production," World Bank Technical Paper, Washington DC, 1988, p. 4.
20. Th. M. Boers and J. Ben-Asher, "A Review of Rainwater Harvesting," *Agricultural Water Management*, Vol. 5, No. 2, 1982, pp. 145-158. http://dx.doi.org/10.1016/0378-3774 (82)90003-8
21. H. J. Finkel and M. Finkel, "Engineering Measures: Water Harvesting," In: H. J. Finkel, M. Finkel and Z. Naveh, Eds., *Semi-Arid Soil and Water Conservation*, CRC Press, Inc., Boca Raton, 1986, pp. 93-101.
22. K. Siegert, "Introduction to Water Harvesting. Some Basic Principles for Planning, Designand Monitoring," In: FAO, *Water Harvesting For Improved Agricultural Production. Expert*

Consultation, Cairo, 21-25 November 1994, pp. 9-23.http://www.fao.org/docrep/u3160e/u3160e03.htm#1.2definitionsandclassification

23. D. Prinz, "Water Harvesting: Past and Future," In: L. S. Pereira, Ed., *Proceedings of Sustainability of Irrigated Agriculture*, NATO Advanced Research Workshop, Rotterdam, 1996, pp. 135-144.
24. S. Zakaria, N. A. Al-Ansari, S. Knutsson and M. EzzAldeen, "Rain Water Harvesting and Supplemental Irrigation at Northern Sinjar Mountain, Iraq," *Journal of Purity, Utility Reaction and Environment*, Vol. 1, No. 3, 2012, pp. 121-141.
25. S. Zakaria, N. A. Al-Ansari, S. Knutsson and M. EzzAldeen, "Rain Water Harvesting at Eastern Sinjar Mountain, Iraq," *Journal of Geoscience Research*, Vol. 3, No. 2, 2012, pp. 100-108.
26. S. Zakaria, N. A. Al-Ansari and S. Knutsson, "Rainwater Harvesting Using Recorded and Hypothetical Rainfall Data Scenarios," *Journal of Earth Sciences and Geotechnical Engineering*, Vol. 3, No. 2, 2013, pp. 21-42.
27. N. A. Al-Ansari, M. Ezz-Aldeen, S. Knutsson and S. Zakaria, "Water Harvesting and Reservoir Optimization in Selected Areas of South Sinjar Mountain, Iraq," *Journal of Hydrologic Engineering*, 2013.
28. J. Anschütz, A. Kome, M. Nederlof, R. de Neef and T. van de Ven, "Water Harvesting and Soil Moisture Retention," 2nd Edition, Stoas Digigrafi, Wageningen, 2003.
29. A. H. Bakari, H. F. Mahoo and N. Hatibu, "Perormance of Maize under Gully Flow Supplementary Irrigation," In: P. J. Makungu, H. F. Mahoo, V. C. K. Silayo and M. Mdemu, Eds., *Proceedings of Tanzania Society of Agricultural Engineers*, Vol. 8, 1998, pp. 20-42.
30. J. W. Gowing, H. F. Mahoo, O. B. Mzirai and N. Hatibu, "Review of Water Harvesting Techniques and Evidence for Their Use in Semi-Arid Tanzania," *Tanzania Journal of Agricultural Science*, Vol. 2, No. 2, 1999, pp. 171-180.
31. O. Mzirai and S. Tumbo, "Macro-Catchment Rainwater Harvesting Systems: Challenges and Opportunities to Access Runoff," *Journal of Animal & Plant Sciences*, Vol. 7, No. 2, 2010, pp. 789-800.

32. H. J. Bruins, M. Evenari and U. Nessler, "Rainwater- Harvesting Agriculture for Food Production in Arid Zones the Challenge of the African Famine," *Applied Geography*, Vol. 6, No. 1, 1986, pp. 13-32. http://dx.doi.org/10.1016/0143-6228(86)90026-3

33. P. Fox and J. Rockstrom, "Supplemental Irrigation for Dry- Spell Mitigation of Rainfed Agriculture in the Sahel," *Agricultural Water Management*, Vol. 61, No. 1, 2003, pp. 29-50. http://dx.doi.org/10.1016/S0378-3774(03)00008-8

34. N. Hatibu, M. D. B. Young, J. W. Gowing, H. F. Mahoo and O. B. Mzirai, "Developing Improved Dryland Cropping Systems for Maize in Semi-Arid Tanzania. Part 1: Experimental Evidence for the Benefits of Rainwater Harvesting," *Experimental Agriculture*, Vol. 39, No. 3, 2003, pp. 279-292. http://dx.doi.org/10.1017/S0014479703001285

35. K. E. Motsi, E. Chuma and B. B. Mukamuri, "Rainwater Harvesting for Sustainable Agriculture in Communal Lands of Zimbabwe," *Physics and Chemistry of the Earth*, Vol. 29, No. 15-18, 2004, pp. 1069-1073.

36. J. Barron and G. Okwach, "Run-Off Water Harvesting for Dry Spell Mitigation in Maize (*Zea mays* L.): Results from On-Farm Research in Semi-Arid Kenya," *Agricultural Water Management*, Vol. 74, No. 1, 2005, pp. 1-21. http://dx.doi.org/10.1016/j.agwat.2004.11.002

37. F. Liu, Y. Wu, H. Xiao and Q. Gao, "Rainwater-Harvesting Agriculture and Water Use Efficiency in Semi-Arid Regions in Gansu Province, China," *Outlook on Agriculture*, Vol. 34, No. 3, 2005, pp. 159-165. http://dx.doi.org/10.5367/000000005774378838

38. K. O. Adekalu, J. A. Balogun, O. B. Aluko, D. A. Okunade, J. W. Gowing and M. O. Faborode, "Runoff Water Harvesting for Dry Spell Mitigation for Cowpea in the Savannah Belt of Nigeria," *Agricultural Water Management*, Vol. 96, No. 11, 2009, pp. 1502-1508. http://dx.doi.org/10.1016/j.agwat.2009.06.005

39. N. Hatibu and H. Mahoo, "Rainwater Harvesting Technologies for Agricultural Production: A Case for Dodoma, Tanzania," In: P. G. Kaumbutho and T. E. Simalenga, Eds., *Conservation Tillage with Animal Traction. A Resource Book of the Animal Traction Network for Eastern and Southern Africa*, ATNESA, Harare, 1999, 173 p.

40. J. Rockstrom, L. Karlberg, S. P. Wani, J. Barron, N. Hatibu, T. Oweis, A. Bruggeman, J. Farahani and Q. Zhu, "Managing Water in Rainfed Agriculture. The Need for a Paradigm Shift," *Agricultural Water Management*, Vol. 97, 2010, pp. 543-550.
41. T. Oweis and A. Hachum, "Water Harvesting for Agriculture in Dry Areas," CRC Press, Balkema, 2012, p. 57. http://dx.doi.org/10.1016/S0378-3774(01)00096-8
42. V. T. Chow, D. R. Maidment and L. W. Mays, "Applied Hydrology," McGraw Hill, New York, 1988, pp. 147- 151.
43. J. R. Williams, "The EPIC Model, Chapter 25," In: V. P. Singh, Ed., *Computer Models of Watershed Hydrology*, Water Resources Publications, Highlands Ranch, 1995.
44. S. L. Neitsch, J. G. Arnold, J. R. Kiniry and J. R. Williams, "Soil and Water Assessments Tool Theoretical Documentation," 2005, p. 102. http://swat.tamu.edu/media/1292/swat2005theory.pdf
45. ASCE, "ASCE/EWRI Curve Number Hydrology Task Committee," In: R. H. Hawkins, T. J. Ward, E. W. Donald and J. A. Van Mullem, Eds., *Curve Number Hydrology*: *State of the Practice*, American Society of Civil Engineers, Reston, 2009, 116 p.
46. H. Mohammed, "Water Harvesting in Erbil Governorate, Kurdistan Region, Iraq. Detection of Suitable Sites Using Geographic Information System and Remote Sensing," Master Thesis, Lund University, 2013, 13 p.
47. P. Buringh, "Soils and Soil Conditions of Iraq," Ministry of Agriculture, Agricultural Research and Projects, Baghdad, 1960.
48. S. Zakaria, N. A. Al-Ansari, Y. Mustafa, S. Knutsson, P. Ahmad and B. Ghafour, "Rainwater Harvesting at Koysinjaq (Koya), Kurdistan Region, Iraq," *Journal of Earth Sciences and Geotechnical Engineering*, 2013.

Chapter 2

Solar Disinfection of Pseudomonas aeruginosa in Harvested Rainwater: A Step towards Potability of Rainwater

Muhammad T. Amin[1*], Mohsin Nawaz[1], Muhammad N. Amin[2], Mooyoung Han[3]

[1]Alamoudi Water Research Chair, King Saud University, Riyadh, Kingdom of Saudi Arabia,

[2]Department of Civil and Environmental Engineering, College of Engineering, King Faisal University, Al-Hofuf Al-Ahsa, Kingdom of Saudi Arabia,

[3]Civil and Environmental Engineering Department, Seoul National University, Shinrimdong, KwanakGu, Seoul, South Korea

ABSTRACT

Efficiency of solar based disinfection of *Pseudomonas aeruginosa (P. aeruginosa)* in rooftop harvested rainwater was evaluated aiming the potability of rainwater. The rainwater samples were exposed to direct sunlight for about 8–9 hours and the effects of water temperature (°C), sunlight irradiance (W/m^2), different rear surfaces of polyethylene terephthalate bottles, variable microbial concentrations, pH and turbidity were observed on *P. aeruginosa* inactivation at different weathers. In simple solar disinfection (SODIS), the complete inactivation of *P. aeruginosa* was obtained only under sunny weather conditions (>50°C and >700 W/m^2) with absorptive rear surface. Solar collector disinfection (SOCODIS) system, used to improve the efficiency of simple SODIS under mild and weak weather, completely inactivated the *P. aeruginosa* by enhancing the disinfection efficiency of about 20% only at mild weather. Both SODIS and SOCODIS systems, however, were found inefficient at weak weather. Different initial concentrations of *P. aeruginosa* and/or *Escherichia coli* had little effects on the disinfection efficiency except for the SODIS with highest initial concentrations. The inactivation of *P. aeruginosa* increased by about 10–15% by lowering the initial pH values from 10 to 3. A high initial turbidity, adjusted by adding kaolin, adversely affected the efficiency of both systems and a decrease, about 15–25%; in inactivation of *P. aeruginosa* was observed. The kinetics of this study was investigated by Geeraerd Model for highlighting the best disinfection system based on reaction rate constant. The unique detailed investigation of *P. aeruginosa* disinfection with sunlight based disinfection systems under different weather conditions and variable parameters will help researchers to understand and further improve the newly invented SOCODIS system.

INTRODUCTION

Water shortage and adequate cost-effective water treatment technologies are the major concerns of this era around the globe. According to the latest available information by the joint study of UNICEF and WHO, still 783 million people do not have access to safe water resources [1]. The use of polluted water for ingestion, especially in many areas of developing countries, is the major cause of diarrhea and various other water-borne diseases. Over 1.2 million children, most being less than five years old, die due to diarrheal diseases each year [2]. Currently the alternate water sources and efficient water treatment technologies are needed. Rainwater harvesting (RWH) as safe water source and cost-effective solar disinfection (SODIS) system has significant importance in this regard [3]–[5]. For efficient utilization of harvested rainwater, the RWH system should consist of its catchment area, a storage tank, a treatment facility, a supply facility and pipes [6]. Roof being an impermeable catchment area is a good source to collect rainwater. However the microbial quality of harvested rainwater can be deteriorated due to the possible contamination upon contact with catchment which is one of the major constraints of using rainwater for potable purposes [7], [8]. *P. aeruginosa* is a part of microbial contamination which is considered responsible for nosocomial infection in immune-compromised people [9]. The presence of this microorganism is unacceptable as it has been implicated in waterborne and food borne diseases and is now considered primary infectious agent [10]. *P. aeruginosa* is detected in stored rainwater by many researchers which is of serious health concerns for water users [11]–[15].

SODIS is being used as point-of use treatment method in many parts of the world. It is believed that solar energy can be used in improving water quality in those parts of the world that experience hot, sunny climates [16], [17]. The portion of solar radiation which plays a major role in generating the antimicrobial activity is believed to be UV-A, 315–400nm, and to a lesser extent visible violet and blue light in the range of 400–490nm [18], [19]. In addition, the increase in water temperature through solar collector is known to work synergistically with solar radiation to further enhance the efficiency of SODIS process [20], [21]. The inactivation of high populations of microorganisms in highly turbid water (approximately 200 nephelometric turbidity units, NTU) can be achieved within 7h if the water temperature reached at least 55°C [22]. Previous studies found that children, between the ages of 5–16 years, who stored their drinking water in 1.5 liters polyethylene terephthalate (PET) bottles in direct sunlight were less prone to diarrhea and other water borne diseases compared to the children who kept their water bottles out of the sun [23], [24]. The efficiency of SODIS and solar collector disinfection (SOCODIS) systems was tested against total and fecal coliform, *Escherichia coli* (*E. coli*) and HPC previously [4] while the efficiency of SODIS and SOCODIS systems was checked against the inactivation of *P. aeruginosa* in this study.

The current study focuses on harvested rainwater as a potential potable water source by using cost-effective solar disinfection of *P. aeruginosa*. The objectives include the enhancement of *P. aeruginosa* inactivation using different rear surface of PET bottles and to investigate the efficiency of solar collector under mild and weak weather conditions with different concentrations of the tested organism. The effects of different initial pH and turbidity values (adjusted by adding kaolin) on *P. aeruginosa* inactivation are also the important part of this study in addition to investigating the effects of the presence of *E. coli* in different concentrations.

MATERIALS AND METHODS

Sampling Sites and Disinfection Setup

The rainwater storage tank was installed in one of the educational buildings (engineering building #39) at the Campus of Seoul National University, South Korea. The harvested rainwater from the rooftop of the building was used mainly for toilet flushing inside building. Rainwater samples were collected in 8 sterile PET bottles of 2L capacity from the supply point which is situated at an height of 1.35 m from the base of the underground rainwater storage tank. The PET bottles were filled ¾ full and were shaken for about 20 seconds to oxygenate the water and an air space of about 15% of the bottle volume was still kept for aeration [16].

In simple SODIS, 3 PET bottles with transparent, absorptive (rear surface painted black) and reflective (rear surface covered with aluminium foil) rear surfaces were exposed to natural sunlight while 4 transparent PET bottles kept inside the box were exposed to sunlight at the same time in SOCODIS system. The detailed design of the SOCODIS system has already been described in a previous study [4]. The actual and schematic diagram of SOCODIS system and reflective SODIS ($SODIS_{(Ref.)}$) in this study is shown as Fig. 1 [25] while the transparent and absorptive SODIS ($SODIS_{(Abs.)}$) cases are not presented. Both SODIS and SOCODIS system was exposed to direct sunlight on the rooftop of the same building where the rooftop RWH system was installed and it was located 37°27′18.31″N, 126°57′3.25″E. There was no need of any special permission to carry out the exposure experiments at the rooftop of the building and the activities did not involve endangered or protected species.

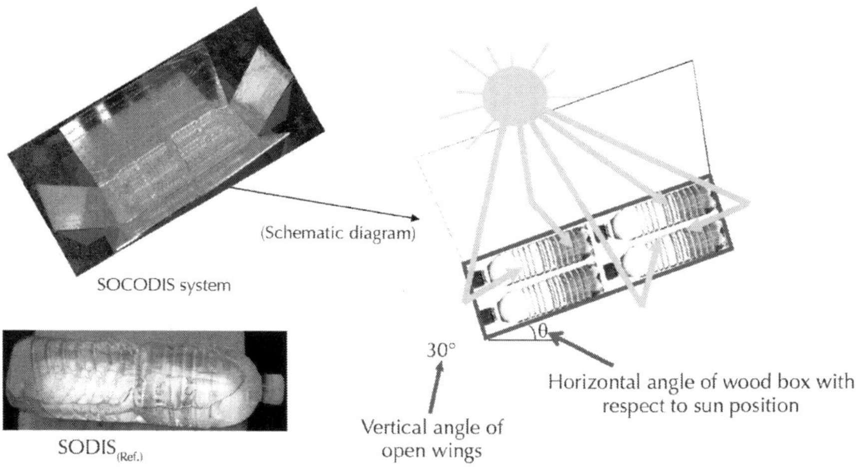

Figure 1: Demonstration of $SODIS_{(Ref.)}$ and SOCODIS systems.

The duration of natural sunlight exposure in both systems, SODIS and SOCODIS was about 9 hours. Experiments were performed at all weathers in case of simple SODIS system but for SOCODIS system, the analysis was made only under mild and weak weather conditions. Sunlight radiations, at all occasions within a year, were monitored on-site with a SP-110 Pyranometer (Apogee Instruments Inc., Logan, USA) connected to a data logger (DT80 Series 2) recording 1 minute averages in Watt/m^2 (W/m^2). The spectral range of the Pyranometer used was 0–1750 W/m^2. The values for irradiance and the temperature inside the bottle were recorded at an interval of an hour.

Microbial Analysis

The microbial water quality analysis was carried out by Standard Methods in Soil and Water Quality Laboratory, Seoul National University (SNU) at room temperature 25°C. The detection of *P.*

aeruginosa was made by multiple tube method. Asparagine broth (Titan Biotech Ltd.) for the presumptive stage of *P. aeruginosa* was used in a series of fifteen test tubes for all three dilutions of 10, 1 and 0.1 ml and thus 5 tubes per each dilution. The tubes were than incubated at 35°C for 24 hours. To check the number of *P. aeruginosa*, the tubes were examined under black light after 24 and then after 48 hours. The tubes with green fluorescent pigments were selected and then 0.1 ml of culture was further incubated at 35°C into acetamide broth (Titan Biotech Ltd.) to complete the confirmation stage [26]. For *E. coli* detection and measurements, DifcoTM Lauryl Tryptose Broth (Becton, DickinsonandCompany) was used for the presumptive phase while the confirmation was done by using the BactoTM EC Medium with Mug (Becton Dickinson France S.A.).

Miscellaneous Measurements

A dark control was separately kept in 1 PET bottle wrapped in aluminium foil, over the same time period for each experiment. Nearly 5–6 repetitions were carried out for most of the analysis to avoid any experimental error and the error bars are shown in almost all the time graphs.

For the effects of different initial pH values, four pH values were chosen for both SODIS and SOCODIS system and the comparison was made only at weak weather conditions. For initial pH adjustment in stored rainwater, diluted HCl and NaOH were used. To check the effect on *P. aeruginosa* inactivation at different turbidity values, kaolin was introduced in raw rainwater samples. The turbidity values were set at 20NTU and 100 NTU and the effects were checked and compared for best suited case $SODIS_{(Abs.)}$ and for SOCODIS system, separately.

The behavior of inactivation curves on experimental data was investigated with GlnaFit (Geeraerd and Van Impe Inactivation Model Fitting Tool) using Geeraerd Log-linear plus shoulder model [27]. Geeraerd model has been widely used in many SODIS scientific contributions to fit experimental results [28].

RESULTS AND DISCUSSIONS

Table 1 shows the average initial values of physicochemical parameters (turbidity, pH, EC & DO) and the number of *P. aeruginosa* observed in stored rainwater samples. The physicochemical parameters were only used as reference values and the major emphasize of this study was on the microbial analysis. The concentration of *P. aeruginosa* in harvested rainwater before disinfection varied throughout the year as 350–900 CFU/100 ml depending upon both wet and dry seasons. No artificial method of *P. aeruginosa* or *E. coli* spiking was used.

Table 1: Reference values of stored rainwater samples

Physicochemical parameters				Microbial parameter		
Turbidity	Temperature	pH	EC	DO	P. aeruginosa	E. coli
NTU	°C		ms/cm	mg/l	CFU/100 ml	CFU/100 ml
2-5	22-26	7-8	150–450	6-9	350-900	200-450

Chemical and microbiological parameters fluctuated during the course of the study, with the highest levels of microbiological contamination in wet seasons when the sunlight irradiance were highest and sunny weather conditions prevailed. Overall, the collected rainwater had a relatively good physicochemical quality but did not meet the requirements for potable water. Most of the water-quality parameters including the bacterial counts were higher both at the beginning and ending of the wet season mainly due to dirty catchment surfaces and addition of first-flush into rainwater tanks after long dry periods. Initial pH was high in dry periods but decreased during the rainy season and increased again after the rains stops. Both *E. coli* and*P. aeruginosa* levels were lower during the wet season due to short water storage time despite the fact that the fresh contamination was added with the rains. Due to filtration of the rainwater through the filters attached to the weirs before entering into the rainwater storage tank, a reduction in the concentrations of *P. aeruginosa* was expected, but this was not always the case. This was evident by observing the microbial load even after the filtrations step and by observing the good numbers of microorganisms even after the first-flush.

Evaluation of Solar Radiation and Temperature under Different Weather Conditions

Solar irradiance plays the most important role in solar based disinfection systems. For efficient SODIS, a geographical area should receive radiation above 500 W/m^2 for 3–5 hours [19], [29].Fig. 2 represents the pattern of sunlight intensity by using the hourly average values of irradiance for categorizing the weather condition into sunny, mild and weak weathers. The time 0h corresponds to 8am or 9am in the morning while 9h corresponds to 5pm or 6 pm in the evening. The irradiance values varied from 700–950 W/m^2 for sunny weather with an average of 775 W/m^2 (usually from May to August), 500–700 W/m^2 for mild weather with an average of 540 W/m^2 (usually from September to October and March to April) and 140–300 W/m^2 with an average of 200 W/m^2 for weak weather (usually from November to February).

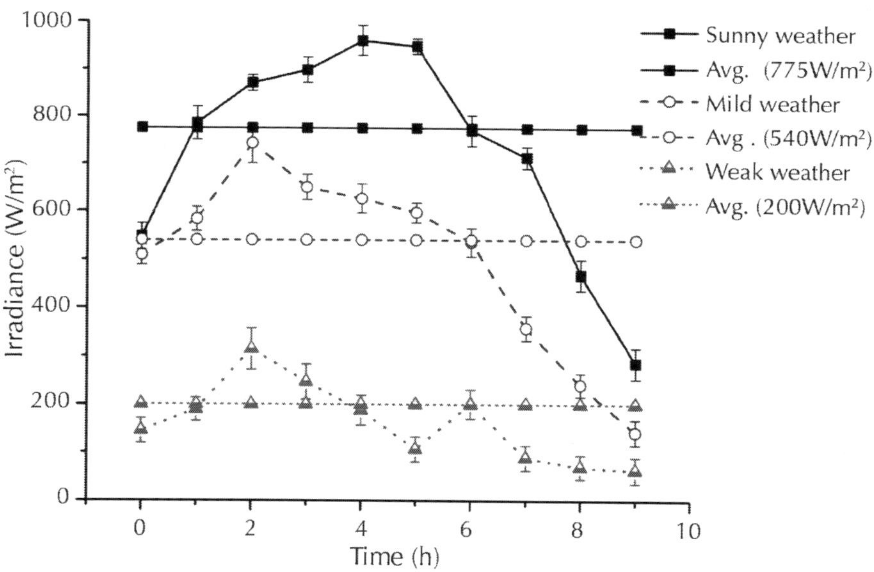

Figure 2: Irradiance change under different weather conditions.

The temperature difference at 9th hours between absorptive and reflective rear surfaces container was 2.5–3°C under all weather conditions. Temperature comparison was also made using the same rear surfaces as mentioned earlier under different weather conditions. The water temperature was measured at regular intervals of 1h by keeping the device inside the bottles while the ambient temperature was not recorded. The average temperature values drawn for sunny, mild and weak weather conditions are shown in Fig. 3(a, b, & c). Absorptive container used the mechanism to enhance the water temperature by absorbing radiation while the irradiance affect was dominant in reflective container.

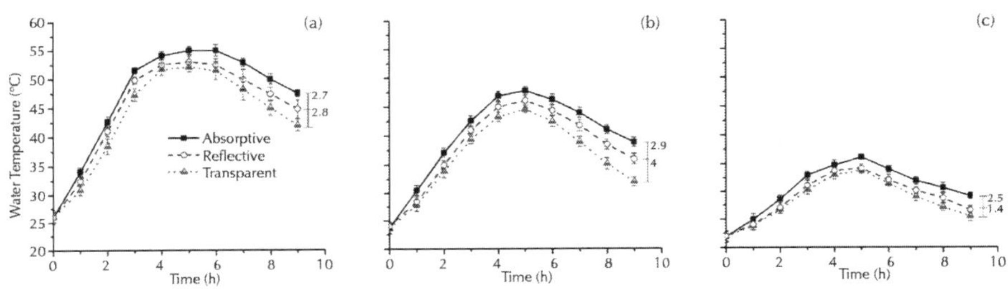

Figure 3: Temperature change in PET bottles with different rear surfaces against exposure time under (a) sunny, (b) mild, and (c) weak weather conditions.

There was almost an exponential rise in water temperature reaching maximum value in the afternoon after which it drops with decreasing sun irradiance. Temperature was affected by weather conditions and was directly proportional to irradiance. This affect can be affirmed by comparing Fig. 2 with Fig. 3. The effective temperature for microbial inactivation was considered above 50°C with dominant thermal effect of sunlight [30]. The results of the current study were comparable with the previous findings as the temperature was raised above 50°C, during sunny weather conditions, in almost all containers shown in Fig. 3(a). However, mild weather conditions may not be fit for SODIS if only thermal rather than synergistic effects of thermal and UV radiations are considered since the maximum temperature attained was less than 50°C. Low temperature under weak sunlight conditions may have adverse effects on the efficiency of SODIS because of less or no thermal and/or synergistic effects of thermal and UV radiations. The observed values of temperature and irradiance were according to the weather pattern of South Korea (Latitude: 37°35′ North, Longitude: 127°03′ East). Similar pattern of temperature and irradiance can be observed for exposure time and can be generalized for any location in the world.

Effects of Different Rear Surfaces of Pet Bottles under Different Weather Conditions in SODIS

The optimization of system by selecting the best type of container suited under different weather conditions was one of the main objectives of this study. For this purpose, transparent, absorptive and reflective containers were used to compare the efficacy of SODIS at sunny, mild and weak weather conditions as shown in Fig. 4 (a, b & c). Fig. 4(a) shows microbial inactivation in all three containers with different rear surfaces under sunny weather conditions. The complete inactivation of *P. aeruginosa* in absorptive container confirmed the dominant effect of thermal inactivation. Also the higher inactivation of *P. aeruginosa* (about 97%) in reflective container showed that synergistic effect of temperature and radiation was available. The inactivation rate under sunny weather condition with different rear surface was in order as abs orptive>reflective>transparent containers. In transparent container the temperature was raised above 50°C under sunny weather condition but the complete inactivation of *P. aeruginosa* was not achieved even after 9h of exposure. This observation showed the resistivity of *P. aeruginosa* to SODIS and was contrary with the previous studies (22, 31, 32]. Under mild weather conditions as shown in Fig. 4(b) however, the inactivation rate of *P. aeruginosa* was unexpectedly higher in reflective containers compared to absorptive containers but the complete inactivation of *P. aeruginosa* was not achieved in any case. The exact reason of this slightly higher inactivation in reflective container was unknown at least in this study.

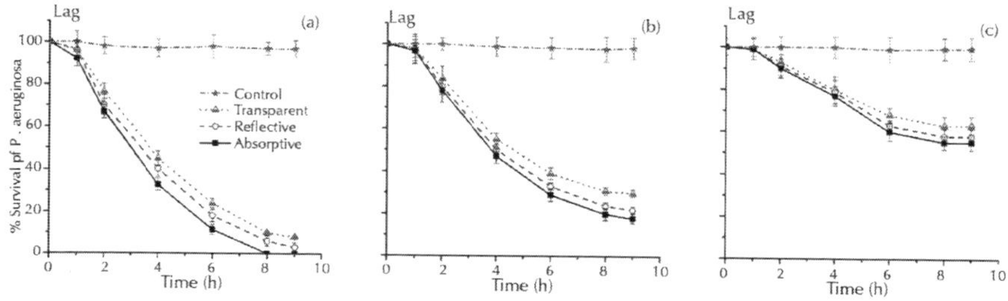

Figure 4: *P. aeruginosa* inactivation with different rear surfaces of PET bottles under (a) sunny, (b) mild, and (c) weak weather conditions.

The difference in inactivation of *P. aeruginosa* with transparent rear surface under mild conditions was about 10%. As none of the water sample, with any rear surface, under mild weather condition could reach to the 50°C so the irradiance, other dominant factor under SODIS, may be the possible reason for *P. aeruginosa* inactivation. The effects of irradiance looked to be more significant with reflective rear surface as indicated in Fig. 4(b). Transparent containers showed poor performance where by both effects of radiation and temperature were not concentrated as was the case in reflective and/or absorptive containers. Weak weather as displayed in Fig. 4(c) showed bad results for similar effects of different rear surfaces. Only 37–45% inactivation of *P. aeruginosa* after 9 hours of exposure to direct sunlight showed that least effect of temperature and/or irradiance was available.

In all above mentioned conditions, except $SODIS_{(Abs.)}$ under sunny weather condition, the inactivation rate of *P. aeruginosa* was almost negligible for first hour. This duration represents the lag phase in which the bacteria showed some resistance to solar inactivation. After lag phase, exponential inactivation of *P. aeruginosa* was observed for further 3–5 hours which was the peak hours of sunlight irradiation on any day and under all weather conditions. For last 2–3 hours of exposure, there was almost negligible inactivation at least under weak sunlight intensity.

Comparison of $SODIS_{(Ref.)}$ and SOCODIS Systems for Radiation and Temperature Effects on Microbial Inactivation

The efficiency of SODIS and SOCODIS systems was investigated under different weather conditions based upon the irradiance range and temperature. During 9 hours exposure to sunlight, the inactivation rate of *P. aeruginosa* was initially measured after 1h and then after every 2h interval. Finally the comparison of SODIS($_{Ref.)}$ was made with SOCODIS system. All the results exhibited almost the same trend observing a close correlation between irradiance and time required to inactivate *P. aeruginosa*. At first, for the comparison purposes of $SODIS_{(Ref.)}$ and SOCODIS, the temperature changes under mild and weak weather conditions were measured at regular intervals of 1h. The average temperature values of all four bottles, kept inside the solar collector, for mild and weak weather conditions are shown in Figs. 5(a) and 5(b), respectively. Comparing the maximum temperature reached in SODIS and SOCODIS system, difference between peak values under mild and weak weather conditions was about 2–3°C. The relatively greater temperature increase in SOCODIS system was observed because the solar irradiation was reflected back onto the PET bottles from all four sides due to the open side wings of solar box. This effect was achieved only from one side i. e. base of the PET bottle, in SODIS $_{(Ref.)}$ case. The synergistic effects of temperature and irradiance were also reported in earlier studies [31] which were evident under mild weather conditions as shown in Fig. 5(c) where the complete inactivation of *P. aeruginosa* was achieved.

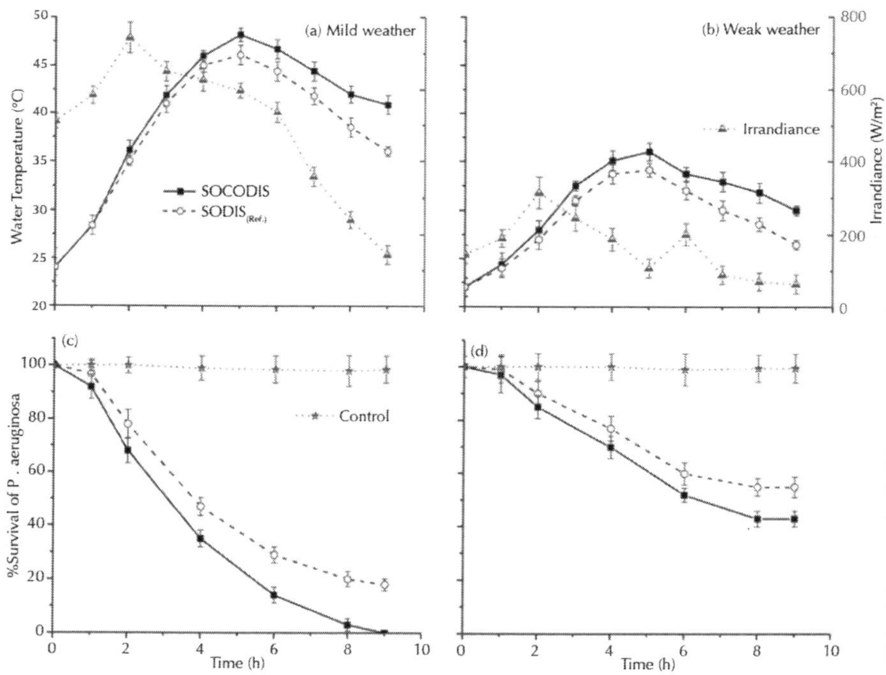

Figure 5: Comparison of temperature and *P. aeruginosa* inactivation in SODIS$_{(Ref.)}$ and SOCODIS system under mild (a & c), and weak (b & d) weather conditions.

However, the SOCODIS system remained inefficient under weak weather conditions, mostly because the synergistic effect of radiation and temperature was absent shown in Fig. 5(d). The difference in disinfection efficiency between the two systems can be summarized as follows: the SOCODIS system was about 20–30% more efficient than the simple SODIS system. It is well known previously that the effectiveness of SODIS improves when a solar collector is used, enhancing the irradiance effect by reflecting and returning the solar energy to the water samples, as support for bottles [31].

The exact mode of action by which SODIS or SOCODIS systems achieved *P. aeruginosa* inactivation is not fully understood. It has recently been demonstrated that solar photons attack proteins in bacteria, directly or indirectly via reactive oxygen species (ROS), causing protein oxidation. This leads to the disruption of metabolic function and damage to the cell membrane losing the bacterial cultivability [33]. *P. aeruginosa* inactivation is directly related to the sunlight intensity and UV dose, especially in the UV-A range. According to some previous studies, after getting the required UV dose, the disruption of bacterial cells continues even the exposed water samples are removed from sun and stored in dark conditions [33], [34].

UV-A is highest as a proportion of solar irradiance when the sun is at its highest altitude near solar noon, as is clear from an increase in the inactivation rate at this time, from 11 am to 3 pm in conducted experiments. All the results showed same tendency signifying a close relationship between UV dose and time required to inactivate microorganisms. One suspicion on solar disinfection by using PET bottles is the leaching of carcinogenic degradation agents from plastic bottles [35]. However, this suspicion was removed in another sunlight exposure study, PET degradation products such as terephthalate monomers and dimers are primarily formed at the outer surface of the bottles [36]. Some compounds including the carbonyls and plasticizers are found in the water but within the safe limits set for drinking water quality [37].

Effects of Different Initial Concentrations of *P. aeruginosa* on Disinfection Efficiency

Since the initial concentrations of *P. aeruginosa* varied throughout the year depending upon the catchment conditions which in turn is influenced by the wet and dry seasons, therefore, the effects of this different initial concentrations on the disinfection efficiency was investigated at mild weather and the results are presented in Fig. 6 for both SODIS and SOCODIS system. Three different initial concentrations of *P. aeruginosa* (350, 700 and 980 CFU/ml) were chosen at different times during the four-months span of each weak and mild weathers but mainly at the beginning and ending of the both weather conditions.

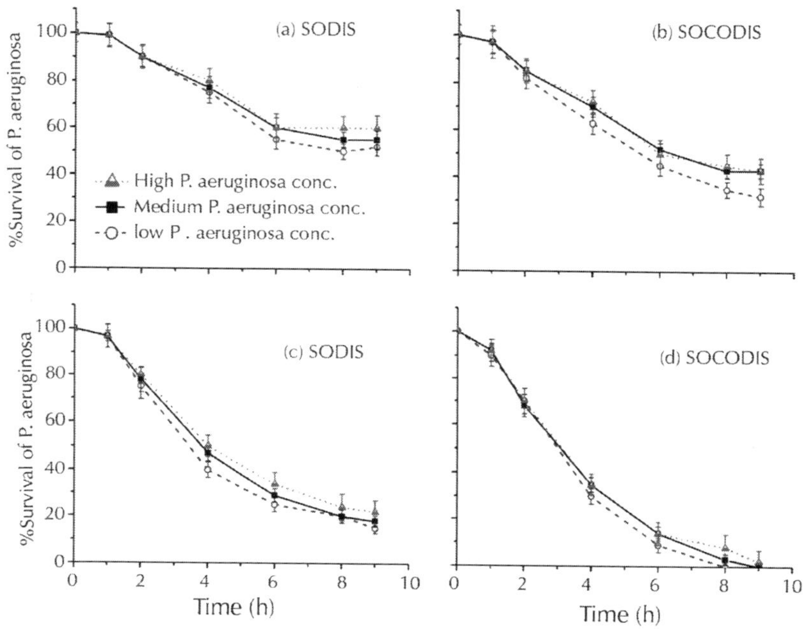

Figure 6: Effects of different initial *P. aeruginosa* concentratinos on disinfection efficiency in both SODIS and SOCODIS system under weak (a & b) and mild (c & d) weather conditions.

Low to high concentrations of *P. aeruginosa* were possible under weak weather conditions due to both wet and dry season for four months from November to February. Mild weather conditions spanning nearly over four month (March, April, September and October) yielded medium to high concentrations of *P. aeruginosa* during March and April, due to relatively dirty catchment surface as a result of dry seasons, and low to medium concentrations, during relatively wet months and resulted clean catchment surface, in September and October. As shown in Fig. 6, there is very little effect of higher initial *P. aeruginosa* concentrations on disinfection efficiency especially in SODIS under both weak and mild weather conditions. The inactivation of *P. aeruginosa*, however, increased slightly about 5–10% by lowering the initial concentrations in SOCODIS system and this could be due to the concentrated effects of sunlight radiations.

Effects of the Presence of *E. coli* on Disinfection of *P. aeruginosa*

The effects of the presence of other bacterial organisms like *E. coli* on disinfection efficiency of *P. aeruginosa* was investigated and for this purpose, three different initial concentrations of *E. coli* (210, 320 and 450 CFU/100 ml as low, medium and high, respectively) were chosen under mild weather conditions. It can be concluded from the results presented in Fig. 7 that the presence of other bacterium like *E. coli* had little effects on the disinfection performance of SODIS or SOCODIS system for *P. aeruginosa*. Higher *E. coli* concentrations, however, slightly decreased the inactivation of *P. aeruginosa* in addition to longer lag period, as is shown in Fig. 7 and this might be due to the increasing competition of different organisms for absorbing sunlight radiations or due to the bacterium selectivity of the UV-A radiations especially during early hours of exposure.

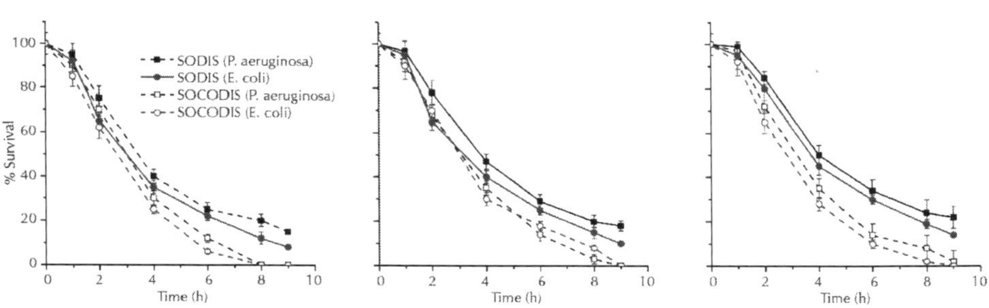

Figure 7: Effects on *P. aeruginosa* inactivation in both SODIS and SOCODIS system at mild weather at (a) low, (b) medium, and (c) high concentrations of *E. coli*.

Reaction Kinetics

Inactivation kinetics of *P. aeruginosa* was investigated by using the Geeraerd Log linear and shoulder model in GInaFit and inactivation rate constant (k_{max}) in all cases of SODIS and SOCODIS systems was calculated and compared. The model was originally defined by coupling two differential equations [27].

$$\frac{dN}{dt} = -k_{max}.N.(\frac{1}{1+C_c}).(1-\frac{N_{res}}{N}) \qquad (1)$$

$$\frac{dC_c}{dt} = -k_{max}.C_c \qquad (2)$$

Where C_c is related to the physiological state of the cells, k_{max} is the specific inactivation rate [1/time unit] and N_{res} is the residual population density [CFU/mL]. Following format of the model is used when actually applying to the experimental data in GInaFit.

$$\log 10(N) = \log 10[(10^{\log 10(N(0))} - 10^{\log 10(Nres)}).e^{-k_{max}t}.\left(\frac{e^{k_{max}S_l}}{1+(e^{k_{max}S_l}-1).e^{-k_{max}t}}\right)+10^{\log 10(N_{res})}] \qquad (3)$$

The first two mentioned equations (eq.1&2) can be obtained by substituting $S_{L,}$ a parameter represents the shoulder (time unit), or N_{res} equal to zero. Table 2 presents a brief overview of*P. aeruginosa* inactivation based on the inactivation rate constants (k_{max}), calculated using the Geeraerd model.

Table 2: Comparison of k_{max} (1/min) for *P. aeruginosa* disinfection between SODIS and SOCODIS system under different weather conditions (Rainwater with low turbidity (2-5NTU) and neutral pH values)

Weather conditions	SODIS			SOCODIS
	Transparent	Absorptive	Reflective	
Sunny	0.47	0.63	0.56	ND*
Mild	0.17	0.22	0.26	0.71
Weak	0.07	0.1	0.11	0.26

The k_{max} (1/min) values in Tables 2 and 3 were calculated based on the Geeraerd model, as described earlier, and represent the inactivation rate constants for *P. aeruginosa*, with a corresponding coefficient of determination (R^2) of 1, except where mentioned. A complete application of the model, i.e. with tailing and shoulder, was not observed in most of the cases and modeling with tailing was unlikely for the observed data probably due to the synergistic effects of UV radiations and temperature. For mild and weak weather conditions, temperature less than 50°C but high k_{max} values validated the assumption that irradiance effects were significant in case of SOCODIS unlike SODIS for the inactivation of *P. aeruginosa*. This finding is in compliance with previously reported results [38].

Table 3: Comparison of k_{max} (1/min) for *P. aeruginosa* disinfection between $SODIS_{(Abs.)}$ (suny weather) and SOCODIS system (mild weather) at different pH and turbidity values

pH (under Turbidity, 5NTU)	Sunny weather	Mild weather	Turbidity (under pH,7)	Sunny weather	Mild weather
	$SODIS_{(Abs.)}$	**SOCODIS**	**NTU**	**$SODIS_{(Abs.)}$**	**SOCODIS**
3	0.83	0.98	~ 5	0.63	0.71
5	0.72	0.83	20	0.48	0.55
7	0.63	0.71	50	0.39	0.36
10	0.59	0.62	100	0.28	0.22

Effects of Different pH and Turbidity Values on Disinfection Efficiency

Both pH and turbidity of the stored rainwater are among the main factors that may change the efficacy of solar disinfection systems. So, the effects of different initial pH and turbidity values on the efficiency of the solar disinfection systems were tested and the results are presented in Figs. 8 and 9, respectively. As mentioned earlier, diluted HCl and NaOH were used for adjusting initial pH values while kaolin was added for different initial turbidity values. As shown in Fig. 8, four initial pH values i.e. 10, 7, 5 and 3 were selected with rainwater with a low initial turbidity value (<5 NTU) and the comparison was performed only under weak weather conditions mainly due to inefficiency of the SOCODIS system at this weather. The overall disinfection efficiency increased by about 15% by lowering the initial pH value from 10 to 3 and nearly complete inactivation of *P. aeruginosa* was achieved at the acidic conditions. Low pH may have increased the inactivation of *P. aeruginosa* by presenting significant additional stress to the cells [5], [39].

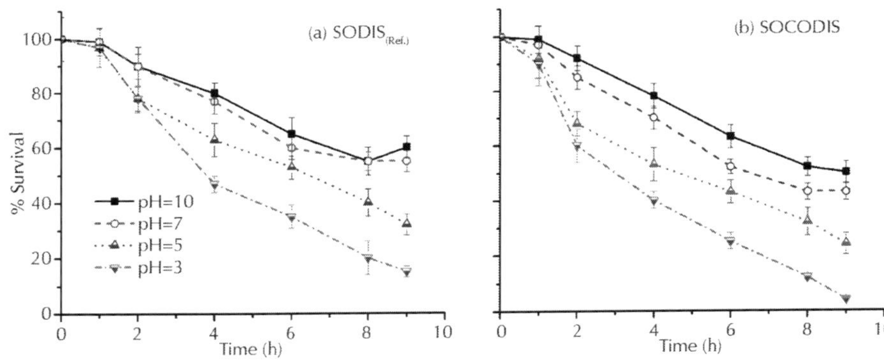

Figure 8: Effects of different initial pH values on *P. aeruginosa* inactivation in (a) $SODIS_{(Ref.)}$, and (b) SOCODIS system at weak weather conditions.

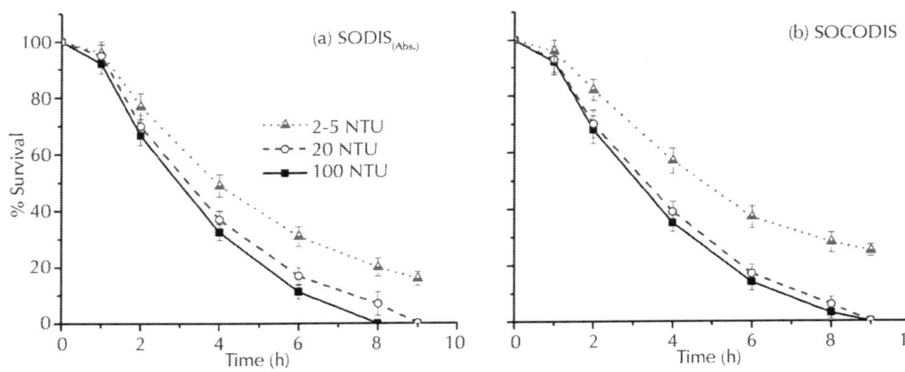

Figure 9: Effects of different turbidity values on *P. aeruginosa* inactivation in (a) $SODIS_{(Abs.)}$ (sunny weather), and (b) SOCODIS system (mild weather).

Table 3 shows a comparison of the *P. aeruginosa* inactivation between SODIS and SOCODIS system under mild and sunny weather conditions at different initial pH values. These findings differed from the earlier studies stating no effects of pH values on microbial inactivation rates[40]. In this study lowering initial pH values had significant effect of *P. aeruginosa* inactivation as reported previously [41]. It is worth mentioning that there are no health-based guidelines for pH and it is reported that the consumption of food or beverages with low (2.5) or high (11) pH does not result in adverse health effects [42].

For investigating the effects of initial turbidity values with rainwater having nearly neutral pH values, best suited solar disinfection cases, SODIS($_{Abs.)}$ under sunny weather and SOCODIS system under mild weather condition were selected for comparison as shown in Figs. 9(a) and 9(b), respectively. Rainwater samples collected from an underground storage tank had a turbidity value of less than 5 NTU. For comparison purposes, higher values (20 and 100NTU) were achieved by adding kaolin. Samples with low turbidity showed better results; the difference between inactivation of *P. aeruginosa* was insignificant up to about 20 NTU. However, the SOCODIS system showed poor performance in terms of *P. aeruginosa*in activation at higher turbidity, and disinfection efficacy decreased by almost 10–15%. This could be because of a loss of UV radiations due to scattering and absorption by suspended particles in rainwater samples. Various microbiological studies have revealed that highly turbid waters do not allow the synergistic effect to occur [22], [31], [38], [43]. However, the highly turbid water may undergo a substantial temperature increase because of the absorption of solar radiation by suspended particles, which may contribute to higher disinfection efficiencies [22]. The case of SODIS($_{Abs.)}$ system under sunny weather conditions was in accordance with previously reported literature but contrarily the complete inactivation of *P. aeruginosa* was not achieved.

Table 3 summarizes the change in the inactivation rate constants of *P. aeruginosa* with the variable values of turbidity when Geeraerd model was used for best suited cases of SODIS($_{Abs.)}$ and SOCODIS systems. As can be noted from the Table, the values of k_{max} were decreased with the increased turbidity level in all cases. Surprisingly, however, the difference between the k_{max} values for turbidity level 20 and 100 under SODIS$_{(Abs.)}$ was lesser (0.20) as compared with SOCODIS case (0.33). It can be due to the substantial temperature increase because of the absorption of solar irradiance by suspended particles [22] when the rear surfaces of PET bottles were painted black.

CONCLUSION

$SODIS_{(Abs.)}$ under, sunny weather conditions, was the most efficient case (100% inactivation) as compared to the $SODIS_{(Trans.)}$ (92%) and $SODIS_{(Ref.)}$ (97%) cases for the inactivation of *P. aeruginosa* in harvested rainwater. This could be due to the thermal and/or synergistic effect of temperature and irradiation. Generally, a 20% increase in the disinfection efficiency of the SOCODIS system, even under mild weather conditions, suggested that the simple solar collector/box is an efficient, cost effective and user friendly option especially for remote areas. However under weak weather conditions, a longer exposure time or other supplementary means with the SOCODIS system are required for the complete inactivation of *P. aeruginosa*.

There was very little effect of high initial *P. aeruginosa* or *E. coli* concentrations on disinfection efficiency but the inactivation of *P. aeruginosa* increased slightly about 5–10% by lowering the initial concentrations in SOCODIS system. Higher *E. coli* concentrations, on the other hand, slightly decreased the inactivation of *P. aeruginosa* in addition to longer lag period. Although, not completely inactivated under weak weather conditions, the overall disinfection efficiency increased by about 15% by lowering the initial pH value from 10 to 3 especially in SOCODIS system. Above 20 up to 100NTU, 16–25% decrease in activation of *P. aeruginosa* suggested that water should be filtered before applying solar disinfection. Overall, the inactivation of *P. aeruginosa* in harvested rainwater by solar disinfection systems can be considered a useful addition in making the stored rainwater available for potable purposes. The detailed investigation and disinfection of *P. aeruginosa* with sunlight based disinfection system under different weather conditions and variable parameters will help researchers to understand and further improve the innovative SOCODIS system. This can also be viewed as an easy approach for the poor communities living in the developing world in accessing the potable water using efficient and cost-effective point-of-use treatment methods.

REFERENCES

1. WHO/UNICEF Joint Monitoring Program for Water Supply and Sanitation (2012) Progress on Drinking Water and Sanitation

Update, USA.

2. Black RE, Cousens S, Johnson HL, Lawn JE, Rudan I, et al. (2010) Child Health Epidemiology Reference Group of WHO and UNICEF Global, regional, and national causes of child mortality in 2008: a systematic analysis. The Lancet 375: 1969–1987. doi: 10.1016/s0140-6736(10)60549-1
3. Amin MT, Han MY (2007) Scope/need of soft path water resource management in developing countries. Water Sci Technol: Water Supp 7(5–6): 185–192.
4. Amin MT, Han MY (2009) Roof-harvested rainwater for potable purposes: application of solar disinfection (SODIS) and limitations. Water Sci Technol 60(2): 419–431. doi: 10.2166/wst.2009.347
5. Amin MT, Han MY (2009) Roof-harvested rainwater for potable purposes: application of solar collector disinfection (SOCODIS). Water Res 43: 5225–5235. doi: 10.1016/j.watres.2009.08.041
6. Han MY, Mun JS (2008) Particle behavior consideration to maximize the settling capacity of rainwater storage tanks. Water Sci Technol 56 (11): 73–79. doi: 10.2166/wst.2007.778
7. Miguntanna NS, Egodawatta P, Goonetilleke A (2010) Pollutant characteristics on roof surfaces for evaluation as a storm water harvesting catchment. Desalin Water Treat 19 (1–3): 205–211. doi: 10.5004/dwt.2010.1886
8. Nawaz M, Han MY, Kim TI, Manzoor U, Amin MT (2012) Silver disinfection of Pseudomonas aeruginosa and *E. coli* in rooftop harvested rainwater for potable purposes. Sci Total Environ 431: 20–25. doi: 10.1016/j.scitotenv.2012.05.022
9. Kerr KG, Snelling AM (2009) Pseudomonas aeruginosa: a formidable andever-present adversary. J Hosp Infect 73(4): 338–344. doi: 10.1016/j.jhin.2009.04.020
10. Morais PV, Mesquita C, Andrade J, DaCosta M (1997) Investigation of persistent colonization by Pseudomonas aeruginosa-like strains in a spring water bottling plant. Appl Environ Microbiol 63(3): 851–856.
11. Uba BN, Aghogho O (2000) Rainwater quality from different roof catchments in the Port Harcourt district, Rivers State, Nigeria. J Water Supply Res T 49: 281–288.

12. Albrechtsen HJ (2002) Microbiological investigations of rainwater and graywater collected for toilet flushing. Water Sci Technol 46: 311–316.
13. Kaushik R, Balasubramanian R (2012) Assessment of bacterial pathogens in fresh rainwater and airborne particulate matter using Real-Time PCR. Atmos Environ 46: 131–139. doi: 10.1016/j.atmosenv.2011.10.013
14. Kaushik R, Balasubramanian R, de la Cruz AA (2012) Influence of air quality on the composition of microbial pathogens in fresh rainwater. Appl Environ Microbiol 78: 2813–2818. doi: 10.1128/aem.07695-11
15. Vialle C, Sablayrolles C, Lovera M, Huau M-C, Jacob S, et al. (2012) Water Quality Monitoring and Hydraulic Evaluation of a Household Roof Runoff Harvesting System in France. Water Resour Manag 26: 2233–2241. doi: 10.1007/s11269-012-0012-6
16. Reed RH (1997) Solar inactivation of fecal bacteria in water: the critical role of oxygen. Lett Appl Microbiol 24(4): 276–80. doi: 10.1046/j.1472-765x.1997.00130.x
17. Sommer B, Marino A, Solarte Y, Salas ML, Dierolf C, et al. (1997) SODIS-an emerging water treatment process. J Water Supply Res T 46(3): 127–137.
18. Acra A, Raffoul Z, Karahagopian Y (1984) Solar disinfection of drinking water and oral rehydration solutions, UNICEF Guidelines for Household Application in Developing Countries.
19. Wegelin M, Canonica S, Mechsner K, Fleischmann T, Pesara F, et al. (1994) Solar water disinfection: scope of the process and analysis of radiation experiments. J Water Supply Res T 43(3): 154–169.
20. Rijal GK, Fujioka RS (2001) Synergistic effect of solar radiation and solar heating to disinfect drinking water sources. Water Sci Technol 43(12): 155–162.
21. Abdel Dayem AM, El-Ghetany HH, El-Taweel GE, Kamel MM (2011) Thermal performance and biological evaluation of solar water disinfection systems using parabolic trough collectors. Desalin Water Treat 36 (1–3): 119–128. doi: 10.5004/dwt.2011.2227

22. Joyce TM, McGuigan KG, Elmore-Meegan M, Conroy RM (1996) Inactivation of fecal bacteria in drinking water by solar heating. Appl Environ Microbiol 62(2): 399–402. doi: 10.1046/j.1365-2672.1998.00455.x
23. Conroy RM, Elmore-Meegan M, Joyce TM, McGuigan KG, Barnes J (1996) Solar disinfection of drinking water and incidence of diarrhea in Maasai children: a controlled field trial. The Lancet 348: 1696–1697. doi: 10.1016/s0140-6736(96)02309-4
24. Hobbins M, Mausezah ID, Tanner M (2003) Home-based drinking water purification: the SODIS Health Study. J Water Supply Res T 56 (4): 245–252.
25. Amin MT, Han MY (2011) Improvement of solar based rainwater disinfection by using lemon and vinegar as catalysts. Desalination 276(1-3): 416–424. doi: 10.1016/j.desal.2011.03.076
26. American Public Health Association, American Water Works Association and Water Environment Federation (2012) Standard methods for the examination of water and wastewater. American Public Health Association, Washington, D.C.
27. Geeraerd AH, Valdramidis VP, Van Impe JF (2005) GInaFiT, a freeware tool to assess non-log-linear microbial survivor curves. Int J Food Microbiol 102: 95–105. doi: 10.1016/j.ijfoodmicro.2004.11.038
28. Geeraerd AH, Herremans CH, Van Impe JF (2000) Structural model requirements to describe microbial inactivation during a mild heat treatment. Int J Food Microbiol 59(3): 185–209. doi: 10.1016/s0168-1605(00)00362-7
29. Oates PM, Shanahan P, Polz MF (2003) Solar disinfection (SODIS): simulation of solar radiation for global assessment and application for point-of-use water treatment in Haiti. Water Res 37: 47–54. doi: 10.1016/s0043-1354(02)00241-5
30. El Ghetany HH, Abdel Dayem A (2010) Numerical simulation and experimental validation of a controlled flow solar water disinfection system. Desalin Water Treat 20(1–3): 11–21. doi: 10.5004/dwt.2010.1207
31. McGuigan KG, Joyce TM, Conroy RM, Gillespie JB, Elmore-Meegan M (1998) Solar disinfection of drinking water contained in transparent plastic bottles: characterizing the bacterial

inactivation process. Appl Environ Microbiol 84(6): 1138–1148. doi: 10.1046/j.1365-2672.1998.00455.x

32. Lonnen J, Kilvington S, Kehoe SC, Al-Touati F McGuigan KG (2005) Solar and photocatalytic disinfection of protozoan, fungal and bacterial microbes in drinking water. Water Res 39: 877–883. doi: 10.1016/j.watres.2004.11.023
33. Bosshard F, Riedel K, Schneider T, Geiser C, Bucheli M, et al. (2010) Protein oxidation and aggregation in UVA-irradiated *Escherichia coli* cells as signs of accelerated cellular senescence. Environ Microbiol 12(11): 2931–2945. doi: 10.1111/j.1462-2920.2010.02268.x
34. Ubomba-Jaswa E, Navntoft C, Polo-Lopez MI, Fernandez-Ibanez P, Mcguigan KG (2009) Solar disinfection of drinking water (SODIS): an investigation of the effect of UV-A dose on inactivation efficiency. Photoch Photobio Sci 8: 587–595. doi: 10.1039/b816593a
35. Lilya D (2001) Analysis and risk assessment of organic chemical migration from reused PET plastic bottles, M.Sc. thesis, College of Graduate Studies, University of Idaho, Moscow.
36. Wegelin M, Canonica S, Alder A, Marazuela D, Suter M, et al. (2001) Does sunlight change the material and content of polyethylene terephthalate (PET) bottles?. J Water Supply Res T 50(3): 125–133.
37. Schmid P, Kohler M, Meierhofer R, Luzi S, Wegelin M (2008) Does the reuse of PET bottles during solar water disinfection pose a health risk due to the migration of plasticisers and other chemicals into the water?. Water Res 42: 5054–5060. doi: 10.1016/j.watres.2008.09.025
38. Kehoe SC, Joyce TM, Ibrahim P, Gillespie JB, Shahar RA, et al. (2001) Effect of agitation, turbidity, aluminum foil reflectors and container volume on the inactivation efficiency of batch-process solar disinfectors. Water Res 35(4): 1061–1065. doi: 10.1016/s0043-1354(00)00353-5
39. Foegeding EA, Lanier TC, Hultin HO (1996) Characteristics of edible muscle tissues. In: Owen Fennema R (Ed.). Food Chemistry. Chapter 15, vol. 3: 879–942.

40. Rincon A, Pulgarin C (2004) Effect of pH, inorganic ions, organic matter and H_2O_2 on*E. coli* k12 photocatalytic inactivation by TiO_2 – implications in solar water disinfection. Appl Catal B Environ 51 (4): 283–302. doi: 10.1016/j.apcatb.2004.03.007
41. Fisher MB, Keenan CR, Nelson KL, Voelker BM (2008) Speeding up solar disinfection (SODIS): effects of hydrogen peroxide, temperature, pH, and copper plus ascorbate on the photoinactivation of *E. coli.* J Water Health 6 (1): 35–51. doi: 10.2166/wh.2007.005
42. NHMRC (1996) Australian Drinking Water Guidelines. National Health and Medical Research Council. Commonwealth of Australia Sydney, Australia.
43. Gomez-Couso H, Fontan-Sainz M, Ares-Mazas E (2010) Thermal contribution to the inactivation of Cryptosporidium in plastic bottles during solar water disinfection procedures. Am J Trop Med Hyg 82(1): 35–39. doi: 10.4269/ajtmh.2010.09-0284

Chapter 3

Microbial Quality and Phylogenetic Diversity of Fresh Rainwater and Tropical Freshwater Reservoir

Rajni Kaushik[1,2], Rajasekhar Balasubramanian[1,2*], Hugh Dunstan[3]

[1]Singapore-Delft Water Alliance, National University of Singapore, Singapore, Singapore, [2]Department of Civil and Environmental Engineering, National University of Singapore, Singapore, Singapore,

[3]School of Environmental and Life Sciences, The University of Newcastle, Callahan, NSW, Australia

ABSTRACT

The impact of rainwater on the microbial quality of a tropical freshwater reservoir through atmospheric wet deposition of microorganisms was studied for the first time. Reservoir water samples were collected at four different sampling points and rainwater samples were collected in the immediate vicinity of the reservoir sites for a period of four months (January to April, 2012) during the Northeast monsoon period. Microbial quality of all fresh rainwater and reservoir water samples was assessed based on the counts for the microbial indicators: *Escherichia coli (E. coli)*, total coliforms, and *Enterococci* along with total heterotrophic plate counts (HPC). The taxonomic richness and phylogenetic relationship of the freshwater reservoir with those of the fresh rainwater were also assessed using 16 S rRNA gene clone library construction. The levels of *E. coli* were found to be in the range of 0 CFU/100 mL – 75 CFU/100 mL for the rainwater, and were 10–94 CFU/100 mL for the reservoir water. The sampling sites that were influenced by highway traffic emissions showed the maximum counts for all the bacterial indicators assessed. There was no significant increase in the bacterial abundances observed in the reservoir water immediately following rainfall. However, the composite fresh rainwater and reservoir water samples exhibited broad phylogenetic diversity, including sequences representing *Betaproteobacteria, Alphaproteobacteria, Gammaproteobacteria, Actinobacteria, Lentisphaerae* and *Bacteriodetes*. Members of the*Betaproteobacteria* group were the most dominant in both fresh rainwater and reservoir water, followed by *Alphaproteobacteria, Sphingobacteria, Actinobacteria* and *Gammaproteobacteria*.

INTRODUCTION

Airborne microorgansims can be transferred to aquatic systems through atmospheric fallout of coarse particles (dry deposition) [1] and rainfall (wet deposition) [2]–[3], leading to changes in the microbial composition of receiving water bodies [4]–[5]. The presence of bacterial pathogens in airborne particulate matter (PM) is of particular concern from the public health perspective as these aerosolized bacteria can form new cells in PM [6] and be metabolically active with the potential to biogeochemically mediate atmospheric chemistry [7]–[9]. The transfer of PM containing viable bacterial pathogens from the atmosphere to water bodies can affect not only ecotoxicology, but also human health through various exposure pathways.

Pathogenic episodes in lakes and reservoirs are often associated with rain events and riverine inflows [4], [10]. Excessive rainfall has been reported to be a significant contributor to historical waterborne disease outbreaks due to mobilization and transport of bacterial pathogens[10]–[12]. Stormwater runoff is a major cause of deterioration of surface water quality in urban areas. When rainfall occurs on paved surfaces, large volumes of water are swiftly carried to drains and discharged into receiving surface waters. Thus, the transport of microbial pollution from rain runoff to lakes and reservoirs is a major concern for management of natural waters worldwide [13]. Changes in the abundance of heterotrophic and coliform bacteria resident in stored water bodies have recently been reported in relation to incoming bacterial loads following rain events [5]. Hence, it is very important to understand the impact of rainfall carrying the live microbial aerosols on the quality of freshwater reservoirs in terms of changes in the abundance of pathogens, the microbial community composition and diversity.

Water treatment organizations and environmental protection agencies frequently use the presence of bacterial indicator organisms (e.g. fecal coliforms and enterococci) and their abundance in surface waters as a surrogate for the risk of contamination by actual pathogenic microorganisms [14]–[16]. Most of the studies reported in the literature characterized some bacteria residing in bulk water at various points in the drinking water supply system using cultivation-based approaches (14,17–18]. In general, heterotrophic plate counts (HPC) are used to assess the overall bacterial quality of drinking water, or natural waters [19]. Furthermore, most of the bacterial cells in natural communities are present in a viable but nonculturable (VBNC) state and therefore are non-culturable by current cultivation methods[20]–[21].

The real composition and dynamics of bacterial communities in freshwater resources remain largely unknown. In particular, limited studies have been conducted to determine the levels of bacterial pathogens in fresh rainwater prior to its collection and storage in order to assess its possible impact on the quality of roof-harvested rainwater and also that of aquatic systems upon deposition [1], [9]. The microbial quality of roof harvested rainwater is not only influenced by that of fresh rainwater, but also by the types of rooftop surfaces, the level of water tanks cleanliness, and the presence of insects' and birds' feces.

However, no systematic study has been reported in the literature so far on the phylogenetic diversity of fresh rainwater describing the taxonomic richness of readily culturable organisms present in this water medium although such studies have been conducted for roof-harvested rainwater [22]. Therefore, there is a strong need for quantifying the levels of bacterial pathogens and phylogenetic diversity in both fresh rainwater and reservoir water following their simultaneous collection at the same sampling site of the reservoir.

Molecular approaches-based methods are good tools to elucidate composition of microbial communities residing in various aquatic environments [23]–[24]. These approaches exploit the use of rRNA as taxonomic marker for microorganisms [25]–[26]. The 16 S rRNA gene is tailor-made for microbial identification due to its universal presence in bacteria, extreme species sequence conservation and evolution-induced interspecies variability [27]. With the recent development and application of large-scale high throughput pyrosequencing-based method[28], community-wide spatial and temporal information on microbial community functional structure and potential activity can be rapidly obtained. Although the pyrosequencing-based approach is able to identify new sequences, it suffers from very high sensitivity to random sampling errors, dominant populations and contaminated non-target DNA [29]. The 16 S rRNA gene clone library method is a relatively old method that has been widely used in bacterial community analyses in various environments [30]–[32]. This method has been successfully applied to understand the role of freshwater bacterioplankton in global biogeochemical processes in the aquatic ecosystems [33]–[34]. However, the 16 S rRNA application to fresh rainwater has not been reported to date. The study of phylogenetic diversity in fresh rainwater and reservoir water could provide a basic understanding of the complex communities of environmental bacteria present in fresh rainwater and their impact on the composition of microbial communities in reservoir water intended for human consumption. The objectives of this work were to (1) study the impact of microbial loading of fresh rainwater on the water quality of a tropical reservoir in Singapore (characterized by heavy rainfall) upon deposition using total bacteria and traditional bacterial indicators and (2) examine the kind of bacterial diversity that fresh rainwater and reservoir water harbor using traditional 16 s rRNA cloning and sequencing method.

MATERIALS AND METHODS

Sampling

Singapore's climate is characterized by uniform temperature and pressure, high humidity, and abundant rainfall. Singapore receives an annual rainfall of about 2400 mm. There are no distinct wet, or dry seasons as rainfall occurs every month of the year. The two main seasons, based on the prevailing dominant winds, are the Northeast monsoon season (from late November to March), and the Southwest monsoon season (from late May to September). April to early May and October to early November are generally the transitional months separating the monsoons. December is usually the wettest month with an average rainfall of 280 mm.

Both rainwater samples and reservoir samples were collected at a tropical reservoir (coordinates-1°22′03″N 103°48′07″E) as part of the Singapore-Delft Water Alliance project. There was no need for any special permission to carry out the water sampling, and the activities did not involve endangered or protected species. Reservoir water samples were collected from four different sites at the reservoir using a grab sampling method along with rainwater samples. Both types of water samples were collected concurrently at the same four sites (No. 1, 2, 3 and 4). The reservoir has a capacity of 27.8 million of water over 304 hectares of water surface. Site 1 was located at the centre of the reservoir with no land use. Site 2 was located at the corner of the reservoir with influences from a local drainage system. Site 3 was situated near a major highway and site 4 was near a golf course, having anthropogenic influence from this land use type (Fig.1).

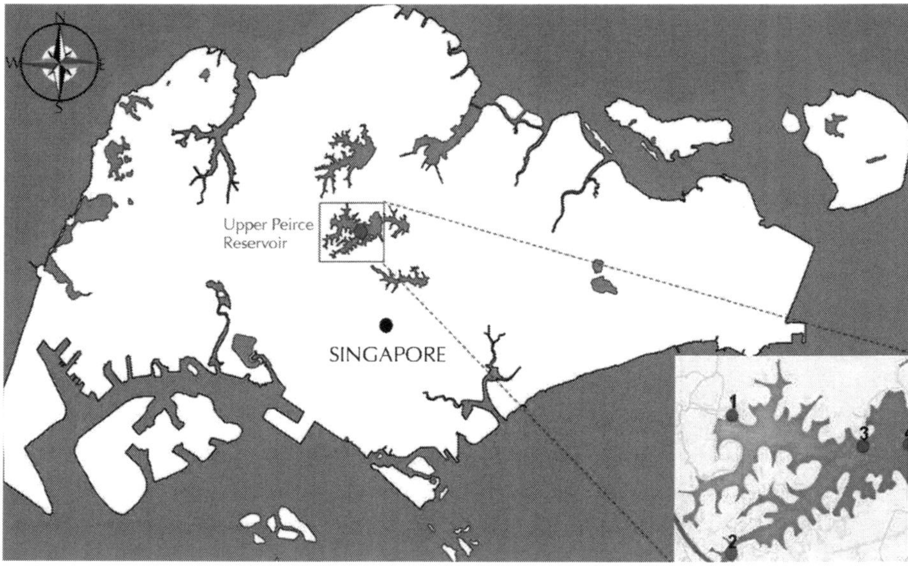

Figure 1: Outline of the four sampling sites at the tropical reservoir of Singapore used in this study.

Both rainwater and reservoir water samples were collected from the sampling locations at regular weekly intervals and within 24 h of rainfall events between January 2012 and April 2012. A total of 33 rain samples on an event-to-event basis were collected (n = 6, 5, 12, and 10 for January, February, March and April, respectively). On each sampling occasion two replicate grab samples (1 L) were collected 1 m from the shore at a depth of ~0.5 m in sterile containers. 1 Liter of fresh rainwater samples was collected in sterilized glass bottles (amber colored glass bottles with a sterilized funnel attached to them) placed at all the four sites used for reservoir sampling. Both fresh rainwater and reservoir water samples (immediately after rainfall) collected from all the four sites were transferred to 2 L sterilized bottles for microbiological and chemical analyses. These water samples were transported to the laboratory in a chilled-cold box and processed within 6 h of collection. The samples were first analyzed for physico-chemical analyses; pH, conductivity, turbidity and conductivity were measured for all the samples as provided in Text S1,Table S1.

Microbiological Analyses

Concentrations of total coliform bacteria and *E. coli* were determined using the m-ColiBlue24 membrane filtration system (Millipore, Cat #M00PMCB24, Bedford, Massachusetts). For all the collected water samples, 100 mL of the sample was filtered onto cellulose ester membranes using vacuum filtration and the membrane was then incubated for 24 h in sterile petri dishes containing absorbent pads soaked with 2 mL of m-ColiBlue24 broth at 37°C. This was performed in duplicate for all the samples. The colonies in blue color were indicative of *E. coli*, while total coliforms were enumerated by counting the colonies in red color. The average CFU/100 mL values obtained for rain events in each month for *E. coli* and total coliforms were estimated.

HPC and *Enterococci* counts were also determined based on average counts obtained for all the samples collected at the four sites in this study. Briefly, for HPC enumeration, one mL of each water sample was serially diluted and the dilutions were aseptically plated in duplicates onto a plate count agar (Sigma-Aldrich, USA) and incubated at 37°C for a maximum duration of 48 h. The average colony counts were expressed as CFU/mL.

For *Enterococci*, the enumeration was performed as per the USEPA Method [35]. In brief, 100 mL of the water sample was filtered onto cellulose esters membranes using vacuum filtration in duplicate and the membranes were then placed on top of the membrane-*Enterococcus* Indoxyl-β-D-Glucoside Agar (mEI Agar, BD, NJ, USA) incubated for 24 h at 41±0.5°C. Colonies with a blue halo, regardless of color, were enumerated as *Enterococci*. The colony counts were expressed as average CFU/100 mL.

DNA Extraction

One liter of each fresh rainwater and reservoir water samples (duplicates for each sampling site and date) was first filtered through a pre-combusted glass fiber filter (90-mm diameter, GF/F, Whatman) followed by a 0.22 μm hydrophilic polycarbonate filter (Millipore Corporation, Bedford, MA). The filters were cut into small pieces and suspended in 1.0 mL of phosphate buffered saline and DNA was extracted using Ultraclean Microbial DNA Isolation kits (MO BIO laboratories, Carlsbad, CA) according to the manufacturer›s instructions. The extracted DNA solution was stored at -80°C.

Construction of 16 S rRNA Gene Libraries

Clone libraries were constructed from composite DNA samples of fresh rainwater and composite DNA samples of reservoir water obtained at the four sites. Briefly, the 16 S rRNA gene was amplified from pooled environmental DNA samples using primers 27f (5′-AGAGTTTGATCMTGGCTCAG-3′) and 1492r (5′-TACCTTGTTACGACTT-3′) [36]. The PCR products were purified using UltraClean PCR Clean-up DNA purification kits (MO BIO Laboratories) and made up to 30 μL with water. Amplified fragments were cloned into TOPO-TA plasmids using the TOPO-TA cloning kit (Invitrogen, Carlsbad, CA) and transferred into*Escherichia coli* DH5a cells (TakaraBio, Otsu, Japan) to construct 16 S rRNA gene libraries.

Cloned plasmid inserts were amplified directly from cells as described [37] using vector primers. The 16 S rRNA gene portion of the cloned DNA was initially sequenced using the ABI Prism BigDye terminator v3.1 and cycle sequencing kit (PE Applied Biosystems). We sequenced 150 clones for each bacterial gene library.

Phylogenetic Analyses

Cloned gene sequences were vector-trimmed and aligned using the NAST (Nearest Alignment Space Termination) algorithm for creating multiple sequence alignments [38]. NAST aligned sequences were chimera checked with chimera slayer. All bioinformatics processes were performed in the MOTHUR environment [39]. Classification of sequences was done using Bayesian classifier method implemented in MOTHUR against Greengenes database [40]. The division level groupings were determined by taxonomic assignment performed by the Ribosomal Database Project 10.0 Classifier tool [41]. Pairwise distances of the aligned clone sequences were calculated and Operational Taxonomic units (OTUs) were grouped using the average neighbor method with a cut-off of 0.03 sequence similarity. Coverage of clone libraries was calculated according to the method recommended by Good (1953) [42]. A Phylogenetic tree was constructed using the neighbor- joining method of ARB with Jukes-Cantor correction model and with 1000 bootstrap replications [43].

Community Statistical Analyses

Microbial quality data were subjected to the student t-test for determining statistical significance. For microbial diversity, analyses of beta diversity with replicate data from rainwater and its respective reservoir water were used for comparison of the two microbial communities. All profiles were inter-compared in a pair-wise fashion to determine a dissimilarity score and stored in a distance matrix. The Unifrac distance metric, as described in Lozupone et al. 2006, was utilized for the phylogenetic distance between OTUs to determine the dissimilarity between the two communities [44]. Weighted Unifrac was used considering the OTU abundance along with the Adonis test of statistical significance ($p<0.05$), which is a non-parametric multivariate analysis of variance (MANOVA) with the Adonis function and utilizes the sample-to-sample distance matrix directly for finding significant differences among these communities [45].

RESULTS AND DISCUSSION

Microbiological Quality of Fresh Rainwater and Reservoir Water

Microbial quality is usually assessed by measuring 'fecal indicator bacteria' (also referred to as fecal indicator organisms, or FIOs) that are generally opportunistic pathogens, present in large numbers in fecal materials. Their presence in water samples is used to indicate the presence of fecal pollution and the possibility that fecal associated pathogens may also be present. The most commonly examined FIOs are thermotolerant coliforms (also termed fecal coliforms),*Escherichia coli* and intestinal *Enterococci* (also termed fecal *streptococci*). *E. coli* are considered the best indicators of fecal contamination in water. The presence of thermotolerant coliforms/*E. coli* in water is unacceptable from the public health perspective as it indicates that a major health risk exists. However, it should be noted that certain members of the coliform group live outside of the gastrointestinal tract in the environment and may create a false indication of fecal contamination [46]. Also, several strains of *Escherichia coli* which are of environmental origin can be thermotolerant, thus giving false positive results [47]. It was therefore necessary to evaluate a range of relevant indicator organisms and total bacteria.

Figs. 2, 3 and 4 show these microbial indicators for both fresh rainwater and reservoir water over the four months of sampling. Levels of thermotolerant coliforms/*E. coli* are expressed as colony forming units (CFU) per 100 mL (CFU/100 mL). The levels of *E. coli* were found to be in the range of 0 CFU/100 mL–75 CFU/100 mL for fresh rainwater and in the range of 10 CFU/100 mL–94 CFU/100 mL in reservoir water (Fig. 2). According to the WHO guidelines for treated drinking water (2006), the levels of *E. coli* should be less than 1 CFU/100 mL [48]. Thus, fresh rainwater is non potable for direct consumption and needs to be treated for potable purposes.

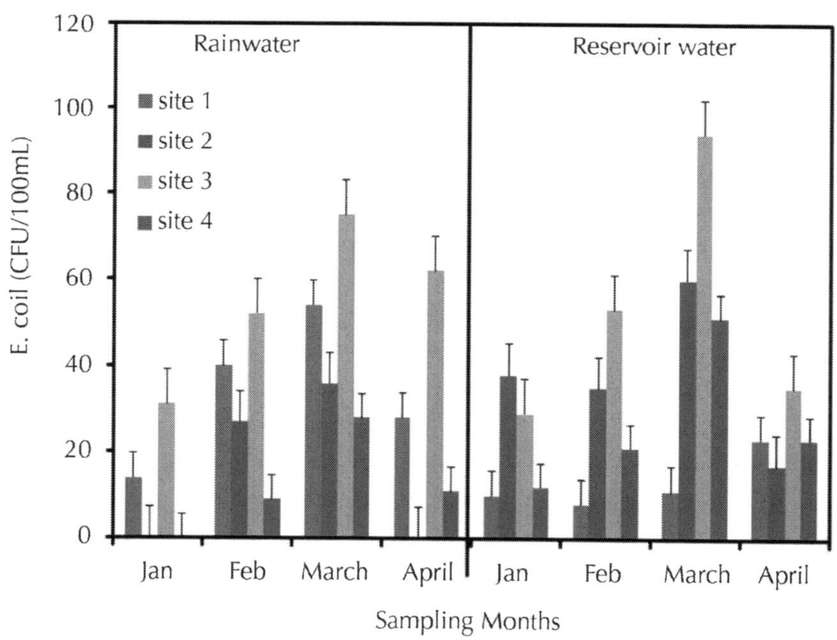

Figure 2: *E. coli* counts from both rainwater samples and reservoir samples collected from four sites from January to April, 2012.

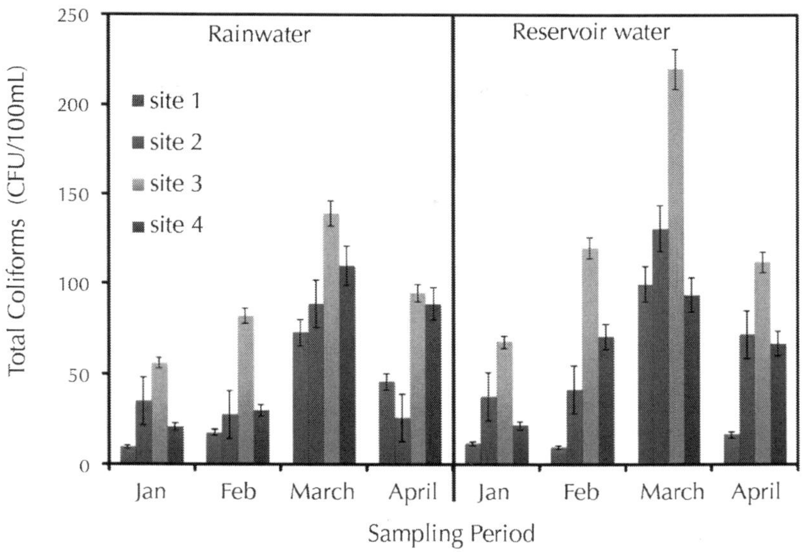

Figure 3: Total coliforms counts from both rainwater samples and reservoir samples collected at four sites from January to April, 2012.

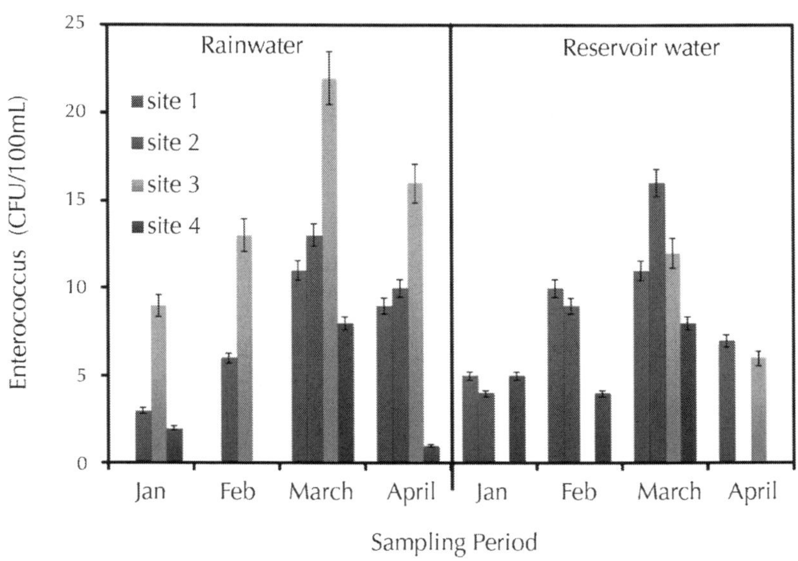

Figure 4: *Enterococcus* counts from both rainwater samples and reservoir samples collected at four sites from January to April, 2012.

Total coliforms were previously considered indicators of fecal contamination. The NHMRC and AWRC "Guidelines for Drinking Water Quality in Australia" (2004) do not consider total coliforms as useful indicators of fecal contamination in the absence of *E. coli*, and have not proposed any guideline value for total coliforms [49]. As the case with HPC, total coliforms can also be used as an indicator for the effectiveness of any treatment program. However, it should be noted that while fecal coliforms are enteric organisms and therefore virtually exclusively of fecal origin, the total coliform count may comprise environmental organisms that are not necessarily common to the digestive tracts of vertebrate animals [50]. The total coliforms were found in the range of 10 CFU/100 mL–220 CFU/100 mL in reservoir water as compared to 10 CFU/100 mL–139 CFU/100 mL in fresh rainwater (Fig. 3). The highest counts were obtained from site #3 in both rainwater and reservoir water, which is near a major highway with residential buildings in the vicinity. This anthropogenic influence could be due to the abundance of airborne microbial pathogens in urban areas with high population density [1], [9]; Turkum *et al.* (2008) reported that most of the species found in rainwater were derived from aerosol and gas-phase components [51].

Enterococci are a specific group of bacteria that are found in high numbers in both human and animal faeces, and are therefore a valuable indicator for determining the extent of fecal contamination of a water source [16], [52]–[53]. *Enterococci* were found to be in the range of 0 CFU/100 mL–11 CFU/100 mL in fresh rainwater and 0 CFU/100 mL–35 CFU/100 mL in the reservoir water (Fig. 4). Overall, the four microbial indicators were found to be the highest at site #3 in both types of water samples. Thus, the results of our microbiological indicator analyses suggested that both the fresh rainwater and the reservoir water samples are not suitable for human consumption without any treatment similar to the findings on the roof harvested rainwater [22], [52], [54].

Total HPC bacteria of the reservoir water samples were two orders of magnitude higher than those of the rainwater. The baseline levels for the reservoir water ranged from 330 CFU/mL to 7.9×10^4 CFU/mL, as compared to 280 CFU/mL to 7.2×10^2 CFU/mL in fresh rainwater (Fig. 5). The reason for the reservoir water having two fold higher magnitudes of total HPC bacteria and slightly higher number of total coliforms could be the additional contribution from sediments apart from rainfall and riverine inflows. HPC bacteria are carried by runoff into the reservoir from highland agricultural areas during intensive rainfall events. Thus, rain events may lead to an inflow of high nutrient concentrations as well as high loads of microbes [55]–[56]. Based on the microbial indicator numbers, there was no statistically significant difference between the rainwater and reservoir water quality with respect to *E. coli* and *Enterococci*. However, the microbial quality of rainwater and reservoir was found to be significantly different ($p \leq 0.05$) for total heterophic counts and total coliforms according to the t-test. It was also observed that the concentrations of all the microbial indicators were higher in the month of March as compared to those in other months. This can be attributed to a higher frequency of rainfall in March as compared to the relatively dry period in the months of January and February.

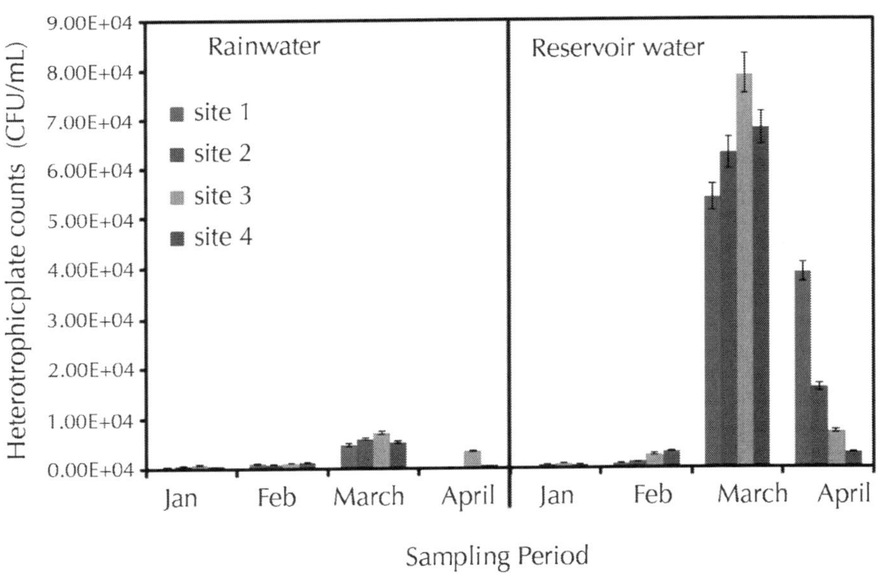

Figure 5: Total heterotrophic plate count bacteria from both rainwater samples and reservoir samples collected at four sites from January to April, 2012.

It should be noted that *E. coli and Enterococci* are indicative of recent fecal contamination as compared to heterophic counts and total coliforms. The influence of *E. coli and Enterococci* of atmospheric origin on the reservoir water quality appears to be relatively less as compared to that of total bacteria HPC and total coliforms; this could due to their lower abundance in the atmosphere and/or the influence of physical processes such as advection, dispersion, diffusion resulting in low residence time [22]. Other processes such as competitive exclusion and nutrient change may also play role to regulate the survival of these incoming bacteria [5].

Microbial Composition and Phylogenetic Analyses of Fresh Rainwater and Reservoir Water

Microbial communities are fundamental to the functioning of aquatic ecosystems. With the availability of metagenomic technologies and approaches [57], it is important to know the changes in total microbial diversity of both rainwater and reservoir water in addition to the changes in microbial indicator numbers. Gaining insights into the total microbial community would provide a better understanding of its role in mediating aquatic ecological processes and biogeochemical cycling [57]–[58]. However, the roles of most microorganisms in natural systems are unclear as most of them cannot be cultivated for investigation by current conventional culturing methods [59]. The introduction of culture-independent molecular methods has shed light on the determination of community compositions, and laid the foundation for gaining a deep understanding of aquatic microbial ecology. Rainwater has received considerable attention as a potential alternative source of potable and non-potable water in regions where there is water scarcity [60]. Rainwater collection is an ancient practice and currently still being practiced, especially in areas with no running water. However, the use of 16 S rRNA sequencing techniques has not been applied to fresh rainwater investigations. Thus, there is a lack of information on its microbial communities and their role in aquatic microbial ecology.

In this study, 10 classes of bacteria were detected in fresh rainwater and four classes of bacteria in reservoir water. In fresh rainwater, sequences were affiliated with *Betaproteobacteria, Alphaproteobacteria, Sphingobacteria, Actinobacteria, Gammaproteobacteria, Lentisphaerae, CH21, Phycisphaerae, Chlorobia* and *Spirochaetes*. In contrast, the reservoir water library detected sequences affiliated with only *Betaproteobacteria, Alphaproteobacteria, Sphingobacteria* and *Gammaproteobacteria*. *Betaproteobacteria* was found to be the dominant class in both the libraries (Fig. 6). The fresh rainwater had higher diversity and taxonomic richness at the class level than those of reservoir water as the numbers of OTUs were found to be 63 and 29 for fresh rainwater and reservoir water, respectively. However, the proportion of *Alphaproteobacteria* in the bacterial community was found to be higher in the reservoir water than in the fresh rainwater. *Alphaproteobacteria*, at least at the class level, are resistant to predation: their relative higher abundance in reservoir water may be due to the capacity to degrade recalcitrant organic compounds such as humic substances accompanied by a tendency of members of the *Alphaproteobacteria* to form filaments, aggregates, or *Caulobacter*-like stalked cells that sometimes can make up a majority of the*Alphaproteobacteria* population in freshwater lakes [61].

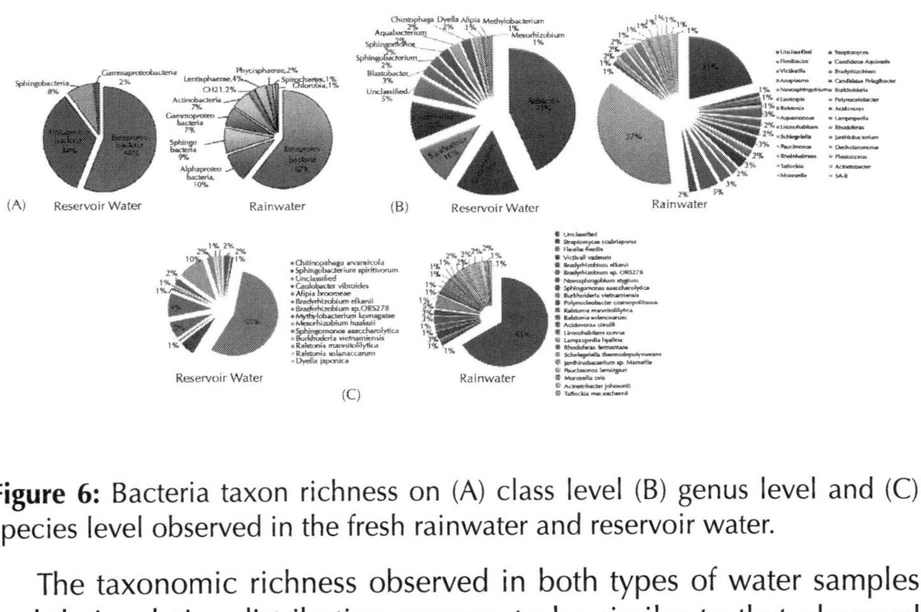

Figure 6: Bacteria taxon richness on (A) class level (B) genus level and (C) species level observed in the fresh rainwater and reservoir water.

The taxonomic richness observed in both types of water samples and their relative distribution appears to be similar to that observed in many freshwater and marine systems, as determined by both cultivation and molecular techniques [22], [62]–[63]; these findings are consistent with the results from previous studies where the dominance of *Betaproteobacteria* has been observed consistently in freshwater systems, particularly among free-living groups, while α and γ sub-classes appear to dominate among particle-attached groups [64]. The high abundance and amenability to culturing have contributed to the *Betaproteobacteria* being the best-studied group in freshwater lakes [61]. Venter *et al.* (2004) carried out a comprehensive genomic study of natural waters at an oligotrophic ocean site in the Sargasso Sea, and found an abundant bacterial distribution dominated by *Proteobacteria* of α, β and γ sub-classes, followed by*Actinobacteria* and *Bacteroidetes* [65]. These findings are similar to those from the fresh rainwater microbial community analyses in the current study as well as from the study conducted by Evans *et al.*, 2009 in rainwater tanks [22]. The class *Betaproteobacteria* was the dominant and the most diverse group in freshwater mesocosm and this dominance is suggested to be associated with their ability to respond quickly to nutrient additions [66]–[67]. The predominance of *Proteobacteria* in fresh rainwater can also be suggestive of the generally clean oligotrophic nature of fresh rainwater.

Actinobacteria are Gram-positive bacteria representing the most abundant group (50–70%) of total bacteria in various freshwater habitats [68]–[69] were also found to be present in fresh rainwater. Previous studies have found that members of *Actinobacteria* contributed to glucose assimilation and heterotrophic nitrification [70]–[71] and played a key role in nutrient and energy cycling in aquatic habits [71]. Also most interestingly, the UV stress resistance has also been postulated to be one of the reasons for the success of the *Actinobacteria* in the upper waters of lakes, which often have high UV transparency. Warnecke and colleagues showed a significant positive relationship between *Actinobacteria* abundance and UV transparency in study on mountain lakes [68]. In addition to UV protection, many members of the *Actinobacteria* are capable of producing spores, allowing them to survive long periods of desiccation. Thus, strong UV protection and desiccation resistance via spore formation, together with the known small cell size of these organisms, would make the freshwater *Actinobacteria* particularly suitable for aerial dispersal, thereby explaining their ubiquitous representation in globally dispersed lakes. However, to date, these taxa have rarely been identified in any air samples [61]. This is the first study to-date to detect their presence in fresh rainwater suggesting their ability to survive in airborne particles that tend to be scavenged by rainwater during rain events [1]. This wet-deposition process enables the global distribution of the microorganisms of atmospheric origin in aquatic ecosystems [66].

Phylogenetic tree-based microbiome comparison of the fresh rainwater and reservoir water samples was performed by the neighbor-joining method of experimentally observed OTUs and reference type strains with each leaf representing an OTU displays counts of sequences observed in each sample (Fig. 7). Notable differences in the frequencies of the some of the bacterial genera were observed. *Bradyrhizobium, Chitinophaga, Caulobacter* and *Ralstonia*were found to be more abundant in reservoir water than in fresh rainwater whereas *Curvibacter*was found to be more abundant in fresh rainwater than in reservoir water. However, unlike taxonomic richness, the coverage of the fresh rainwater clone library (150 sequences from pooled DNA) as calculated based on the species OTU (97% 16 S rRNA gene sequence identity) by Good's Clone Coverage [42] was low (77%) as compared to the coverage of the reservoir water clone library (92%).

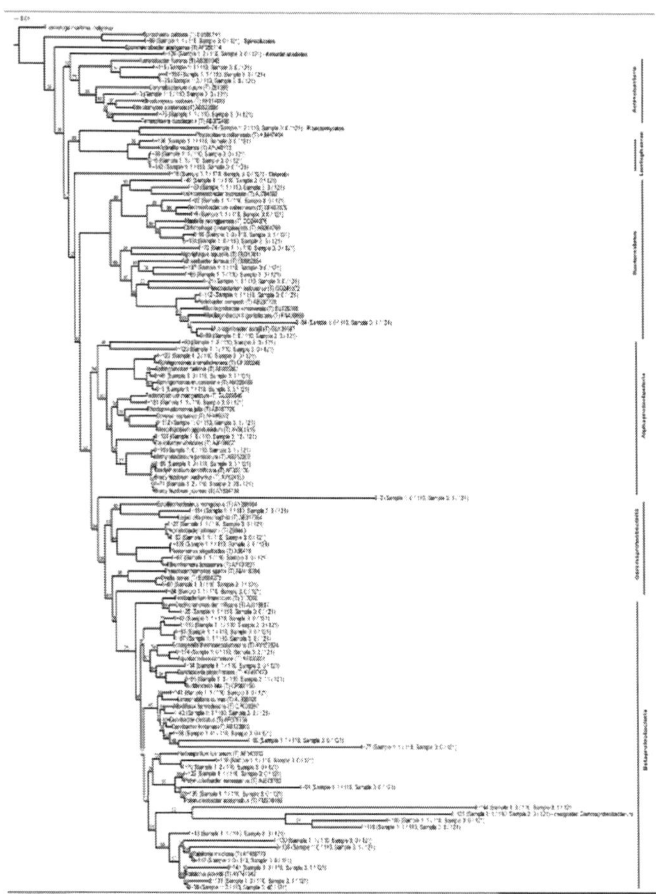

Figure 7: Taxonomic tree indicating the phylogenetic distribution of all bacterial groups in both rainwater and reservoir water samples.

The OTU with the greatest difference in relative abundance was associated with *Curvibacter* for fresh rainwater and with *Ralstonia* for reservoir water. *Curvibacter* and *Ralstonia* belong to the*Burkholderiales* order of *Betaproteobacteria,* under the taxa Comamonadaceae that is reported to be the most abundant typical freshwater fast growing and nutrient-loving group [31], [72]. Rainwater and reservoir samples formed separate groups in ordination analysis, but were not found to be significantly dissimilar (Adonis p-value = 0.125) using OTU abundance metrics. The same diversity of bacteria being present in two different environmental samples suggests that the microbiological composition of the atmosphere and that of the hydrosphere are intimately linked with each other through wet and dry deposition mechanisms. The type of bacterial diversity present in fresh rainwater, the general abundance distribution, and the resemblance of the composition to that of reservoir, have indicated the likely existence of definable micro ecosystems in fresh rainwater and their impact on other aquatic systems. The functional operation of a stable micro-ecology, dominated by well-adapted core resident groups, may have implications with regard to the harvesting of rainwater and for management of water resources.

CONCLUSIONS

This study focused on the investigation of the microbial diversity of fresh rainwater and that of reservoir water with a concurrent assessment of bacterial counts and species profiles in a tropical region, characterized by abundant rainfall throughout the year, for the first time. The results of the study suggested that despite the presence of bacterial pathogens in rainwater and their wet deposition through rainfall, there was no significant increase in the microbial loading of the tropical reservoir water. It appears that the incoming bacterial loads from the rainwater entering the reservoir either undergo dilution, or are not sustained for a long period of time due to the influence of physical processes such as advection, dispersion and diffusion and other processes (competitive exclusion and nutrient change) that may act to regulate the survival of incoming bacteria. Thus, despite the presence of microbial pathogens, rainwater harvesting in large water catchment areas is a promising freshwater resource following treatment. These findings on the composition of aquatic microbial communities in the reservoir water and fresh rainwater by 16 s rRNA clone libraries indicated that the members of*Betaproteobacteria* dominated both the communities. The OTU with the greatest difference in the relative abundance for fresh rainwater was *Curvibacter* whereas for reservoir water it was*Ralstonia*. *Actinobacter* was detected in fresh rainwater, suggesting their presence in airborne particles as well. The fresh rainwater showed greater taxonomic richness than that of the reservoir water.

However, due to lower coverage in traditional cloning method, our understanding about the bacterial composition and phylogenetic diversity of the bacterial community in the freshwater is still limited. Hence, high-throughput molecular tools such as pyrosequencing and hybridization to PhyloChip array that are able to elucidate not only cultivable bacteria, but also viable-but-not-cultivable (VBNC) may extend and expand our understanding of the tropical freshwater reservoir microbial communities. More in-depth studies on the linkage between atmospheric inputs of microorganisms and the microbial diversity of freshwater resources are warranted. This research group has recently performed study of reservoir microbial diversity using PhyloChip microarrays at the same reservoir in the presence and absence of rain events (work in progress), which may add more insights into differences between the traditional cloning and metagenomics of freshwater bacterial diversity.

ACKNOWLEDGMENTS

The authors gratefully acknowledge the support and contributions to Singapore-Delft Water Alliance, National University of Singapore (NUS) for the support provided to the pursuit of this study. Rajni Kaushik thanks SDWA for supporting her Ph.D. study. The technical assistance for sampling provided by Chen Sijing is gratefully acknowledged.

REFERENCES

1. Kaushik R, Balasubramanian R (2012) Assessment of bacterial pathogens in fresh rainwater and airborne particulate matter using Real-Time PCR. Atmos Environ 46: 131–139. doi: 10.1016/j.atmosenv.2011.10.013

2. Womack AM, Bohannan BJM, Green JL (2010) Biodiversity and biogeography of the atmosphere. Philos Trans R Soc Lond B Biol Sci 365: 3645–3653. doi: 10.1098/rstb.2010.0283
3. Kaushik R, Balasubramanian R, de la Cruz AA (2012) Influence of air quality on the composition of microbial pathogens in fresh rainwater. Appl Environ Microbiol 78: 2813–2818. doi: 10.1128/aem.07695-11
4. Brookes JD, Antenucci J, Hipsey M, Burch MD, Ashbolt NJ, et al. (2004) Fate and transport of pathogens in lakes and reservoirs. Environ Int 30: 741–759. doi: 10.1016/j.envint.2003.11.006
5. Martin AR, Coombes PJ, Harrison TL, Dunstan RH (2010) Changes in abundance of heterotrophic and coliform bacteria resident in stored water bodies in relation to incoming bacterial loads following rain events. J Environ Monitor 12: 255–260. doi: 10.1039/b904042k
6. Dimmick RL, Wolochow H, Chatigny MA (1979) Evidence that bacteria can form new cells in airborne particles. Appl. Environ. Microbiol 37: 924–927.
7. Ariya P, Nepotchatykh O, Ignatova O, Amyot M (2002) Microbiological degradation of atmospheric organic compounds. Geophys Res Lett 29: 1–4. doi: 10.1029/2002gl015637
8. Amato P, Ménager M, Sancelme M, Laj P, Mailhot G, et al. (2005) Microbial population in cloud water at the Puy de Dôme: Implications for the chemistry of clouds. Atmos Environ 39: 4143–4153. doi: 10.1016/j.atmosenv.2005.04.002
9. Kaushik R, Balasubramanian R (2013) Discrimination of viable from nonviable Gram-negative bacterial pathogens in airborne particles using propidium monoazide-assisted qPCR. Sci Total Environ 449: 237–243. doi: 10.1016/j.scitotenv.2013.01.065
10. Auld H, MacIver D, Klaassen J (2004) Heavy rainfall and waterborne disease outbreaks: The Walkerton example. J Toxicol Environ Health A 67: 1879–1887. doi: 10.1080/15287390490493475
11. Curriero FC, Patz JA, Rose JB, Lele S (2001) The association between extreme precipitation and waterborne disease outbreaks in the United States, 1948–1994. Am J Public Health 91: 1194–1199. doi: 10.2105/ajph.91.8.1194

12. Ferguson C, Husman AMD, Altavilla N, Deere D, Ashbolt N (2003) Fate and transport of surface water pathogens in watersheds. Crit Rev Env Sci 33: 299–361. doi: 10.1080/10643380390814497
13. Olson BH, Nagy LA (1984) Microbiology of potable water. Adv Appl Microbiol 30: 73–132. doi: 10.1016/s0065-2164(08)70053-4
14. Ward NR, Wolfe RL, Justice CA, Olson BH (1986) The identification of gram-negative, non-fermentative bacteria from water: problems and alternative approaches to identification. Adv Appl Microbiol 31: 293–365. doi: 10.1016/s0065-2164(08)70446-5
15. Edberg SC, Rice EW, Karlin RJ, Allen MJ (2000) *Escherichia coli*: the best biological drinking water indicator for public health protection. J Appl Microbiol 88: 106S–116S. doi: 10.1111/j.1365-2672.2000.tb05338.x
16. Sidhu J, Ahmed W, Toze S (2013) Application of Microbial Source Tracking Toolbox to Identify Sewage Contamination in Stormwater Run-off in Brisbane AWA Water. 40: 81–85.
17. LeChevallier MW, Babcock TM, Lee RG (1987) Examination and characterization of distribution system biofilms. Appl Environ Microbiol 53: 2714–2724.
18. Sartory DP (2004) Heterotrophic plate count monitoring of treated drinking water in the UK: a useful operational tool. Int J Food Microbiol 92: 297–306. doi: 10.1016/j.ijfoodmicro.2003.08.006
19. Oliver JD (2000) The public health significance of viable but nonculturable bacteria. In: R. RColwell and D. JGrimes (ed). Nonculturable microorganisms in the environment. ASM Press, Washington D.C., pp. 277–300.
20. Giovannoni SJ, Britschgi TB, Moyer CL, Field KG (1990) Genetic diversity in Sargasso Sea bacterioplankton. Nature 345: 60–63. doi: 10.1038/345060a0
21. Szewzyk U, Szewzyk R, Manz W, Schleifer KH (2000) Microbiological safety of drinking water. Annu Rev Microbiol 54: 81–127. doi: 10.1146/annurev.micro.54.1.81
22. Evans CA, Coombes PJ, Dunstan RH, Harrison T (2009) Extensive bacterial diversity indicates the potential operation of a dynamic micro-ecology within domestic rainwater storage systems. Sci Total Environ 407: 5206–5215. doi: 10.1016/j.scitotenv.2009.06.009

23. Woese CR (1987) Bacterial evolution. Microbiol Rev 51: 221–271.
24. Höfle MG, Haas H, Dominik K (1999) Seasonal dynamics of bacterioplankton community structure in a eutrophic lake as determined by 5S rRNA analysis. Appl Environ Microbiol 65: 3164–3174.
25. Altmann D, Stief P, Amann R, De Beer D, Chramm A (2003) In situ distribution and activity of nitrifying bacteria in freshwater sediment. Environ Microbiol 5: 798–803. doi: 10.1046/j.1469-2920.2003.00469.x
26. Tringe SG, Hugenholtz P (2008) A renaissance for the pioneering 16S rRNA gene. Curr Opin Microbiol 11: 442–446. doi: 10.1016/j.mib.2008.09.011
27. Schwarzenbach RP, Egli T, Hofstetter TB, von Gunten U, Wehrli B (2010) Global Water pollution and human health. Annu Rev Env Resour 35: 109–136. doi: 10.1146/annurev-environ-100809-125342
28. Huber JA, Mark Welch D, Morrison HG, Huse SM, Neal PR, et al. (2007) Microbial population structures in the deep marine biosphere. Science 318: 97–100. doi: 10.1126/science.1146689
29. Zhou J, Kang S, Schadt CW, Garten CT Jr (2008) Spatial scaling of functional gene diversity across various microbial taxa. Proc Natl Acad Sci USA 105: 7768–7773. doi: 10.1073/pnas.0709016105
30. Hiorns WD, Methe BA, NierzwickiBauer SA, Zehr JP (1997) Bacterial diversity in Adirondack Mountain lakes as revealed by 16 S rRNA gene sequences. Appl Environ Microbiol 63: 2957–2960.
31. Zwart G, Hiorns WD, Methe B, van Agterveld MP, Huismans R, et al. (1998) Nearly identical 16 S rRNA sequences recovered from lakes in North America and Europe indicate the existence of clades of globally distributed freshwater bacteria. Syst Appl Microbiol 21: 546–556. doi: 10.1016/s0723-2020(98)80067-2
32. Wu X, Xi WY, Ye WJ, Yang H (2007) Bacterial community composition of a shallow hypertrophic freshwater lake in China, revealed by 16S rRNA gene sequences. FEMS Microbiol Ecol 61: 85–96. doi: 10.1111/j.1574-6941.2007.00326.x

33. Zehr JP, Ward BB (2002) Nitrogen cycling in the ocean: new perspectives on processes and paradigms. Appl Environ Microbiol 68: 1015–1024. doi: 10.1128/aem.68.3.1015-1024.2002
34. Mason OU, Di Meo-Savoie CA, Van Nostrand JD, Zhou JZ, Fisk MR, et al. (2009) Prokaryotic diversity, distribution, and insights into their role in biogeochemical cycling in marine basalts. ISME J 3: 231–242. doi: 10.1038/ismej.2008.92
35. U.S. Environmental Protection Agency (USEPA) (2002) Method 1600: Enterococci in water by membrane filtration using membrane-Enterococcus Indoxyl-β-D-Glucoside Agar (mEI). EPA-821-R-02-022. Office of Water, Washington, D.C.
36. Weisburg WG, Barns SM, Pelletier DA, Lane DJ (1991) 16S ribosomal DNA amplification for phylogenetic study. J Bacteriol 173: 697–703.
37. Vergin KL, Rappe' MS, Giovannoni SJ (2001) Streamlined method to analyze 16 S rRNA gene clone libraries. Biotechniques 30: 938–940.
38. DeSantis TZ, Hugenholtz P, Keller K, Brodie EL, Larsen N, et al. (2006) NAST: a multiple sequence alignment server for comparative analysis of 16S rRNA genes. Nucleic Acids Res 34: W394–W399. doi: 10.1093/nar/gkl244
39. Schloss PD, Westcott SL, Ryabin T, Hall JR, Hartmann M, et al. (2009) Introducing mothur: Open-Source, Platform-Independent, Community-Supported Software for Describing and Comparing Microbial Communities. Appl Environ Microbiol 75: 7537–7541. doi: 10.1128/aem.01541-09
40. McDonald D, Price MN, Goodrich J, Nawrocki EP, DeSantis TZ, et al. (2012) An improved Greengenes taxonomy with explicit ranks for ecological and evolutionary analyses of bacteria and archaea. ISME J 6: 610–618. doi: 10.1038/ismej.2011.139
41. Cole JR, Wang Q, Cardenas E (2009) The Ribosomal Database Project: improved alignments and new tools for rRNA analysis. Nucleic Acids Res 37: D141–D145. doi: 10.1093/nar/gkn879
42. Good IJ (1953) The Population Frequencies Of Species And The Estimation Of Population Parameters. Biometrika 40: 237–264. doi: 10.1093/biomet/40.3-4.237

43. Ludwig W, Strunk O, Westram R, Richter L, Meier H, et al. (2004) ARB: a software environment for sequence data. Nucleic Acids Res 32: 1363–1371. doi: 10.1093/nar/gkh293
44. Lozupone C, Hamady M, Knight R (2006) UniFrac—an online tool for comparing microbial community diversity in a phylogenetic context. BMC Bioinformatics 7: 371–385.
45. He Z, Xu M, Deng Y, Kang S, Kellogg L, et al. (2010) Metagenomic analysis reveals a marked divergence in the structure of belowground microbial communities at elevated CO2. Ecol Lett 13: 564–575. doi: 10.1111/j.1461-0248.2010.01453.x
46. Pisciotta JM, Rath DF, Stanek PA, Flanery DM, Harwood VJ (2002) Marine bacteria cause false-positive results in the Colilert-18 rapid identification test for Escherichia coli in Florida waters. Appl Environ Microbiol 68: 539–544. doi: 10.1128/aem.68.2.539-544.2002
47. Edberg SC, Allen MJ, Smith DB, Kriz NJ (1990) Enumeration of total coliforms and Escherichia coli from source water by the defined substrate technology. Appl Environ Microbiol 56: 366–369.
48. WHO (2006) Guidelines for drinking-water quality. Geneva: World Health Organization.
49. National Health and Medical Research Council (2004) Guidelines for drinking water quality in Australia. National Health and Medical Research Council (NHMRC & AWRC), Canberra, Australia.
50. Evans CA, Coombes PJ, Dunstan RH (2006) Wind, rain and bacteria: The effect of weather on the microbial composition of roof-harvested rainwater. Water Res 40: 37–44. doi: 10.1016/j.watres.2005.10.034
51. Turkum A, Pekey H, Pekey B, Tuncel G (2008) Investigating relationships between aerosol and rainwater compositions at different locations in Turkey. Atmos Res 89: 315–323. doi: 10.1016/j.atmosres.2008.03.010
52. Ahmed W, Richardson K, Sidhu J, Toze S (2012) Escherichia coli and Enterococcus spp. in rainwater tank samples: comparison of culture-based methods and 23 S rRNA gene quantitative PCR

assays. Environ Sci Technol 46: 11370–11376. doi: 10.1021/es302222b

53. Savichtcheva O, Okabe S (2006) Alternative indicators of fecal pollution: Relations with pathogens and conventional indicators, current methodologies for direct pathogen monitoring and future application perspectives. Water Res 40: 2463–2476. doi: 10.1016/j.watres.2006.04.040
54. Ahmed W, Sidhu J, Toze S (2013) Faecal Indicators and pathogens in Potable Rainwater tanks in Southeast Queensland. AWA Water 40: 88–92.
55. Merz J, Dangol PM, Dhakal MP, Dongol BS, Nakarmi G, et al. (2006) Rainfall–runoff events in a middle mountain catchment of Nepal. J Hydrol 331: 446–58. doi: 10.1016/j.jhydrol.2006.05.030
56. Sargaonkar A (2006) Estimation of land use specific runoff and pollutant concentration for Tapi River Basin in India. Environ Monit Assess 117: 491–503. doi: 10.1007/s10661-006-0769-2
57. Edwards RA, Rohwer F (2005) Viral metagenomics. Nat Rev Microbiol 3: 504–510. doi: 10.1038/nrmicro1163
58. Tseng CH, Chiang PW, Shiah FK, Chen YL, Liou JR, et al. (2013) Microbial and viral metagenomes of a subtropical freshwater reservoir subject to climatic disturbances. ISME J 7(12): 2374–86. doi: 10.1038/ismej.2013.118
59. Amann RI, Ludwig W, Schleifer KH (1995) Phylogenetic identification and in situ detection of individual microbial cells without cultivation. Microbiol Rev 59: 143–169.
60. Meera V, Mansoor Ahammed M (2006) Water quality of rooftop rainwater harvesting systems: a review. J. Water Supply Res Technol.-AQUA 55 (8) 257–267. doi: 10.2166/aqua.2006.052
61. Newton RJ, Jones SE, Eiler A, McMahon KD, Bertilsson S (2011) A guide to the natural history of freshwater lake bacteria. Microbiol Mol Biol Rev 75: 14–49. doi: 10.1128/mmbr.00028-10
62. Newton RJ, Kent AD, Triplett EW, McMahon KD (2006) Microbial community dynamics in a humic lake: differential persistence of common freshwater phylotypes. Environ Microbiol 8: 956–70. doi: 10.1111/j.1462-2920.2005.00979.x
63. Liu Z, Huang S, Sun G, Xu Z, Xu M (2012) Phylogenetic diversity, composition and distribution of bacterioplankton community in

the Dongjiang River, China. FEMS Microb Ecol 80: 30–44. doi: 10.1111/j.1574-6941.2011.01268.x

64. Crump BC, Armbrust EV, Baross JA (1999) Phylogenetic analysis of particle-attached and free-living bacterial communities in the Columbia river, its estuary, and the adjacent coastal ocean. Appl Environ Microbiol 65: 3192–3204.

65. Venter JC, Remington K, Heidelberg JF, Halpern AL, Rusch D, et al. (2004) Environmental Genome Shotgun Sequencing of the Sargasso Sea. Science 304: 66–74. doi: 10.1126/science.1093857

66. Burkert U, Warnecke F, Babenzien D, Zwirnmann E, Pernthaler J (2003) Members of a readily enriched beta-proteobacterial clade are common in surface waters of a humic lake. Appl Environ Microbiol 69: 6550–6559. doi: 10.1128/aem.69.11.6550-6559.2003

67. Simek K, Hornak K, Jezbera J, Masin M, Nedoma J, et al. (2005) Influence of top-down and bottom-up manipulations on the R-BT065 subcluster of β-Proteobacteria an abundant group in bacterioplankton of a freshwater reservior. Appl Environ Microbiol 71: 2381–2390. doi: 10.1128/aem.71.5.2381-2390.2005

68. Warnecke F, Amann R, Pernthaler J (2004) Actinobacterial 16S rRNA genes from freshwater habitats cluster in four distinct lineages. Environ Microbiol 6: 242–253. doi: 10.1111/j.1462-2920.2004.00561.x

69. Lemke MJ, Lienau EK, Rothe J, Pagioro TA, Rosenfeld J, et al. (2009) Description of freshwater bacterial assemblages from the upper Parana River flood pulse system, Brazil. Microb Ecol 57: 94–103. doi: 10.1007/s00248-008-9398-3

70. Brierley EDR, Wood M (2001) Heterotrophic nitrification in an acid forest soil: isolation and characterisation of a nitrifying bacterium. Soil Biol Biochem 33: 1403–1409. doi: 10.1016/s0038-0717(01)00045-1

71. Elifantz H, Malmstrom RR, Cottrell MT, Kirchman DL (2005) Assimilation of polysaccharides and glucose by major bacterial groups in the Delaware Estuary. Appl Environ Microbiol 71: 7799–7805. doi: 10.1128/aem.71.12.7799-7805.2005

72. Glockner FO, Zaichikov E, Belkova N, Denissova L, Pernthaler J, et al. (2000) Comparative 16S rRNA analysis of lake bacterioplankton

reveals globally distributed phylogenetic clusters including an abundant group of Actinobacteria. Appl Environ Microbiol 66: 5053–5065. doi: 10.1128/aem.66.11.5053-5065.2000

Chapter 4

Dryland Soil Hydrological Processes and Their Impacts on the Nitrogen Balance in a Soil-Maize System of a Freeze-Thawing Agricultural Area

Wei Ouyang[1*], Siyang Chen[2], Guanqing Cai[1], Fanghua Hao[1]

[1]School of Environment, State Key Laboratory of Water Environment Simulation, Beijing Normal University, Beijing, China,

[2]Marine Monitoring and Forecasting Center of Zhejiang, Hangzhou, China

ABSTRACT

Understanding the fates of soil hydrological processes and nitrogen (N) is essential for optimizing the water and N in a dryland crop system with the goal of obtaining a maximum yield. Few investigations have addressed the dynamics of dryland N and its association with the soil hydrological process in a freeze-thawing agricultural area. With the daily monitoring of soil water content and acquisition rates at 15, 30, 60 and 90 cm depths, the soil hydrological process with the influence of rainfall was identified. The temporal-vertical soil water storage analysis indicated the local *albic* soil texture provided a stable soil water condition for maize growth with the rainfall as the only water source. Soil storage water averages at 0–20, 20–40 and 40–60 cm were observed to be 490.2, 593.8, and 358 m^3 ha^{-1}, respectively, during the growing season. The evapo-transpiration (ET), rainfall, and water loss analysis demonstrated that these factors increased in same temporal pattern and provided necessary water conditions for maize growth in a short period. The dry weight and N concentration of maize organs (root, leaf, stem, tassel, and grain) demonstrated the N accumulation increased to a peak in the maturity period and that grain had the most N. The maximum N accumulative rate reached about 500 mg $m^{-2}d^{-1}$ in leaves and grain. Over the entire growing season, the soil nitrate N decreased by amounts ranging from 48.9 kg N ha^{-1} to 65.3 kg N ha^{-1} over the 90 cm profile and the loss of ammonia-N ranged from 9.79 to 12.69 kg N ha^{-1}. With soil water loss and N balance calculation, the N usage efficiency (*NUE*) over the 0–90 cm soil profile was 43%. The soil hydrological process due to special *soil* texture and the temporal features of rainfall determined the maize growth in the freeze-thawing agricultural area.

INTRODUCTION

During the agricultural tillage management, the maximum crop production and minimum diffuse nitrogen (N) loading are the priority issues that need to be considered at the same time [1],[2]. It is essential to understand the fates of soil water (SW) and N in agricultural systems in order to attain higher crop yields and N usage efficiency [3]. In developing countries, there is much political and commercial pressure for controlling the N application with the sacrifice of the crop harvest [4]. The dryland agriculture in the Sanjian Plain, Northeast China, is a key food base and the most water limited agricultural zone in China [5]. In this freeze-thawing area, the soil N and water efficiency are the keys for dryland tillage sustainability due to the short growing season. However, there are few reports about the dryland soil hydrological process and N use efficiency in freeze-thawing agricultural areas [6].

The soil nutrient N is a major limiting factor for crop growth and production, which is directly related to the N fertilisation and soil N background level [7]. The soil microbial community, which affects the soil N cycle [8], is less active in this freeze-thawing area. The application of chemical N is a basic agricultural practice in worldwide, which causes excessive discharge of N to the aquatic environment [9]. The European Union agricultural landscape contributes about 55% of the diffuse pollution for water eutrophication. This is a major emerging environmental issue in the developing counties [10]. The soil eco-hydrological process is an essential channel for N transport in the soil-crop system.

Plant organs have different allocation rates with N during growth, and the canopy green area has higher N absorption than the other organs during the jointing period [11]. The N accumulations in the upper leaves of the canopy also follow the eco-hydrological pattern in farmland [12]. Some models have been developed to simulate the N absorption in different organs during crop growth with the impacts of the solar radiation, rainfall, temperature, soil hydrology and N concentration [13]. With the temporal pattern of N accumulation in crop organs, the N cycle and efficiency can also be analyzed. The diffuse N discharge from agricultural systems depends on diverse factors, including the climatic, soil properties and agronomic features [14],[15]. The tillage practice, soil hydrological process and N usage efficiency are the main potential factors to consider when the goal is to reduce the agricultural diffuse N pollution [16]. The dryland maize farmland in a freeze-thawing agricultural area presents different N discharge patterns due to the special hydrological process [17].

The soil hydrological process is the bridge for diffuse N loss from soil to water bodies [18]. The nitrate-N is the dominant type of diffuse N in the soil profile and it can move swiftly downstream flow with the rainfall through the soil pathways [19]. The soil hydrological process is also closely related to the soil texture, which is the medium for water movement and crop growth. In the freeze-thawing dryland, the soil has short-term, transitional phase and rainfall is the only water source for N movement with crop growth [20]. The soil pattern in the study area is meadow*albic* soil (*Albic Luvisols*) and has a 20–30 cm depth impermeable stratum. The *albic* soil provides a stable soil water (SW) condition in the crop root zone and interfaces with the soil hydrological process. With the consideration of high N efficiency, it is necessary to analyze the special pattern of soil hydrology and its impact on the N efficiency [21]. Evaluating the response of N loss to the combination of rainfall and the soil hydrological process can help to identify the interactions of water with N and reduce diffuse N pollution.

The general objective is to maximize the N efficiency in tillage practices. Typically, of about 20%–70% of N is lost from soil-crop systems and this loss can cause other environmental issues [22]. The quality of soil properties will decline under the combination of intensive application of N fertilizer and low N efficiency; this in turn, will also deliver some risk to water [23]. There are several methods to improve the N efficiency, including accurate fertilizer application and irrigation [24],[25]. The dryland soil hydrological process in a freeze-thawing agricultural area presents a special pattern and also affects the N efficiency [26]. This study aims (1) to identify the temporal-vertical dynamics of dryland SW within an *albic* soil layer; (2) to express the dry weight and N accumulation of maize organs in dryland of a freeze-thawing agricultural area; (3) to explore the variation of soil loss, potential N loss, and N use efficiency (*NUE*) in dryland during the growing season.

MATERIALS AND METHODS

Ethics Statement

The field experiment was conducted in the farmland of Bawujiu Farm, which has been rented by Wang Dongli for 50 years. He is also the contact person for future permission. No specific permissions were required for these locations. The field studies did not involve endangered or protected species. This study did not involve vertebrate. The specific location of our study is 133° 50′–134° 33′ E, 47° 18′47°50′ N.

Study Area Description

The case study area was conducted on a farm in the north-eastern part of China. The farm has a long history and on the east is adjoined to Russia

(Fig. 1). The observation site is located in the southern part of the Farm and the elevation is 48 m. The only water source for dryland tillage is precipitation and maize is the dominant crop. This area the temperate continental monsoon climate has an average annual precipitation of 588 mm and an average yearly temperature of 2.94 °C [27]. From October to the following April, the temperature is below zero and, as a result, the soil freeze-thawing process affects the tillage process. The crop growing season is from May to October.

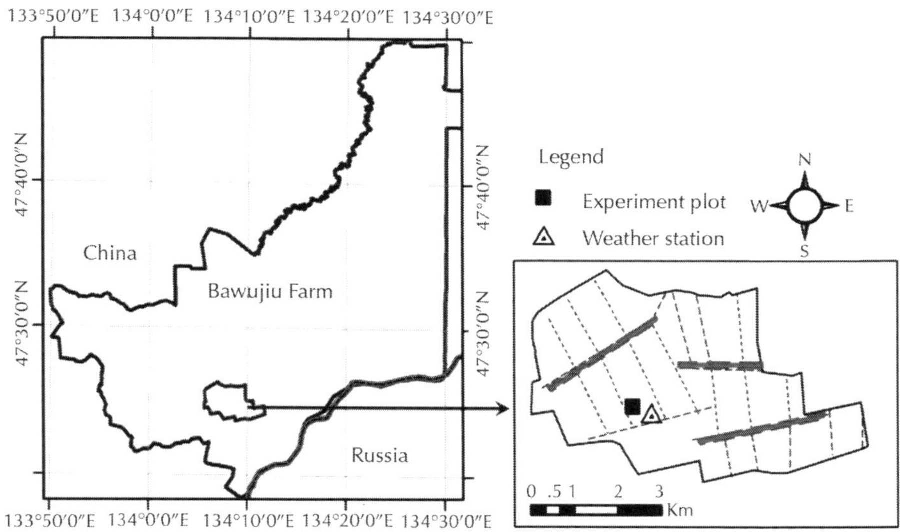

Figure 1: Location of the experimental dryland site in Northeast of China.

Field Monitoring and Soil Sample Collection

In order to identify soil physical-chemical properties, soil samples at four depths (0–15, 15–30, 30–60 and 60–90 cm) were collected in 2010. The samples from three points chosen at random in each layer were mixed into a composite soil sample and placed in plastic bags for lab measurements. The soil samples were collected before sowing and after harvest. Soil particle size distributions were measured by laser diffraction after the removal of organic residue (MasterSizer S, Malvern Instruments, Malvern, UK). The soil's total nitrogen (TN) and organic carbon (OC) concentrations were measured with a CHN Elemental Analyser (Euro VectorS.P.A EA3000, 136 Milan, Italy) [28]. The ammonia N in rainfall was analyzed with the aid of Nessler's reagent and the nitrate-N concentration was determined by the Ultraviolet Spectro Photometric Method. The soil's pH was determined with a pH meter (METTLER TOLEDO, Switzerland) after mixing the fresh soil with deionised water (1:2.5, w/v) [24]. The soil N stock at four depths was calculated with the following equation,

$$\mathrm{Stock_i} = \mathrm{BD_i} * \mathrm{H_i} * \mathrm{C_i} * 10^5 \quad (1)$$

where $Stock_i$ is the soil N storage at layer i (kg/m^2), BD_i is the soil bulk density at layer i(g/cm^3), H_i is the soil depth at layer i (cm), and C_i is the soil N concentration at layer i (mg/kg).

For the soil hydrological process monitoring, the SW collector head (suction cup: Teflon and quartz; OD 21 mm×L 95 mm; porosity: 2 μm; conductivity: 3.31×10^{-6} mm sec^{-1}, PRENART, Danmark) was set at 15, 30, 60 and 90 cm depths. It was covered with a mixed solution of quartz powder and stirred, which can cause a 0.05 MPa vacuum pressure and in direct contact with the soil. The collector tail was connected with PVC pipe and the water samples were collected in bottles by a portable manual vacuum pump. The soil water storage was calculated with soil volumetric water content at different depths. The mean soil water storage at each layer was defined as the average of SW in the upper and lower soil layers. The evapo-transpiration in dryland is calculated with the Bowen ratio and energy balance (BREB). The SW balance was calculated with the following equation [29],

$$\mathrm{SL} = \mathrm{R} - \mathrm{I_r} - \mathrm{ET} - \Delta\mathrm{S} \quad (2)$$

where SL is the soil water loss (mm), R is the precipitation (mm), ET is the evapo-transpiration (mm), I_r is the canopy interception (mm), and ΔS is the soil water storage change (mm).

The local meteorological characteristics were monitored with a ZENO station (Coastal, Seattle, WA, USA). Automated soil temperature sensors (Thermistor, Coastal, Seattle, WA, USA) and soil volumetric water content sensors (TDR type, Coastal, Seattle, WA, USA) were placed in each of the four different soil layers. The monitoring sensors were repeated at double at horizontal 0.5 m space.

Crop Sample Collection and Measurement

The local dryland is intensively cultivated with the aid of mechanized equipment and the maize (*Zea mays L.*) planting density is 75,000 plants ha^{-1}. The detailed maize planting date, harvest date, relevant tillage and fertilization practices are listed in Table 1. The maize samples in 2011 were collected monthly and the sample in 2010 were only collected in October due to the limited experimental conditions. The detailed process of sample collection and monitoring of crop features (height (CH), leaf area index (LAI) and root depth (RD)) are listed in a previously published paper [24]. The maize samples were taken to the lab immediately and the root, leaves, stems, tassels, and grains were separated. The samples were washed clean and then dried for 48 h at 65°C. The dry weight of organ samples was determined after fixing the temperature for 30 min at 105°C. The dried maize samples were crushed and sieved for crop organ N content (%) analysis (Vario EL, Elementar Co. Ltd., Germany). The N uptake amount by crop was calculated based on crop dry weight and plant N content (%).

Table 1: Dryland tillage practices and fertilization information

Crop year	Planting date	Harvest date	Tillage prior to planting	Main fertilisation (kg ha^{-1})
2010	8/Jun.	9/Oct.	Tilling	525
2011	30/May.	5/Oct.		525
Detailed fertilisation data and amounts * (kg ha^{-1})				
2010	2011	N	P	K
8/Jun.	30/May.	71.25	58.65	27.23
27/Jun.	20/Jun.	34.50	13.80	10.02
5/Jul.	1/Jul.	29.45		

Nitrogen Supply and Loss Calculation

The N use efficiency (*NUE*) in a crop-soil system is defined as the ratio of N uptake (N_c) by a crop to N supply (N_s) in a system [30]. The N_s was defined as the total available N to the crop, which included fertilizer N (N_f), mineralized N (N_m), initial mineralized N in soil ($N_{min\ initial}$), N fixed by soil (N_x) and N deposition (N_d) from rainfall. The N_s is calculated as:

$$N_s = N_f + N_{min\ initial} + N_m + N_x + N_d \quad (3)$$

The field N_m is calculated with the crop-soil N equation [31]:

$$N_m + (N_{min\ initial} - N_{min\ final}) = (N_l + N_c + N_{gl}) - (N_f + N_d + N_x) \quad (4)$$

where N_m is the mineralized N, $N_{min\ initial}$ is the initial mineralized N concentration in soil, $N_{min\ final}$ is the final mineralized N concentration in soil, N_l is the soil loss N, N_{gl} is the N loss in gas.N_c is the N crop absorption, N_x is the N fixed by soil, and N_d is the N deposition.

The mineralized N is calculated with the following equation,

$$N_m = (N_l + N_c) - (N_f + N_d) - (N_{min\ initial} - N_{min\ final}) \quad (5)$$

The potential soil water N loss (N_l) is defined as the total loss of nitrate and ammonia, which was estimated with the N concentration in SW and the water volume in dryland [32]. Based on the SW loss and the N concentration in SW at different depths, the soil N loss by water was calculated.

Data Analyses

Data analysis was performed with Sigma plot 10.0 and SPSS 16.0 software. The differences in dry weight and their percentages on crop organs (including root, leaf, stem, tassel, grain) in different growing stages were tested with ANOVA. The difference in the crop N uptake allocation proportion, the N accumulation in different organs, and the mineralized N concentration in different soil layers were statistically analyzed by ANOVA. The multiple comparisons were analyzed with the Duncan method.

RESULTS

Temporal-Spatial Distribution of Soil Water in Dryland

With the soil-maize water monitoring system, the daily precipitation, SW acquisition rate, and SW content in the entire growing period was determined (Fig. 2). The monthly precipitation in the years 2010 and 2011 displayed a similar pattern, except for the lower value in September 2010. The precipitation occurred mainly between July and September, which was also the prime period for maize growth. With the light precipitation in April and May, the maize seed was sowed in moist soil. The precipitation in August and September of 2010 was 64 mm and 120 mm less than in the corresponding months of 2011. The cumulative precipitation from April to September in 2011 was 575 mm, which was 141 mm more than that in 2010. The precipitation was the only water source for the dryland in this freeze-thawing agricultural area, which directly impacted the SW acquisition rate. The SW could be collected only after the rainfall and the vertical difference in the SW acquisition rate was slight (around 0.002 cm3 s−1). In the entire monitoring period, the biggest acquisition rate occurred at the 90 cm depth after the rainfall in early September. The direct reason for it was that only smaller amount of the water can move deeply in the vertical direction after the absorption by roots.

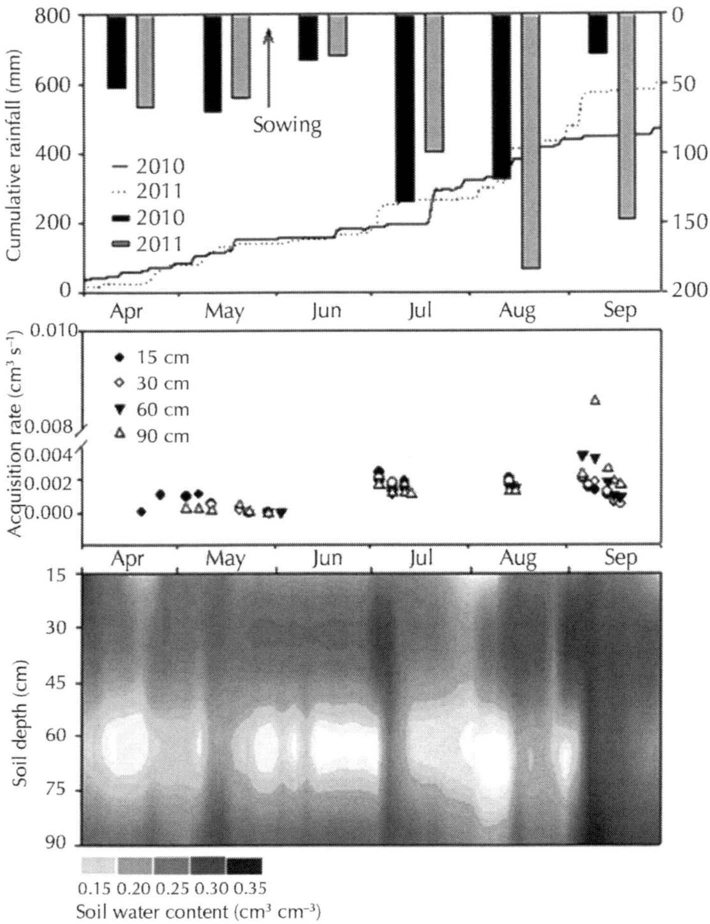

Figure 2: Daily precipitation in 2010 and 2011, and acquisition rate, and soil hydrological pattern of dryland in 2011.

The SW content increased in a stable manner during the growth period and also coincided with the accumulative precipitation. In all of June, the SW content at the 30 cm depth was the highest, the reason for this was that uptake of water in growing roots resulted in water accumulation in the rhizosphere. Due to the 73.6 mm rainfall on July 3, the SW content at the 30 cm depth increased by 20% to 0.30 cm^3 cm^{-3}. In the next two days, the SW content was transported in the vertical direction and it increased to 0.30 cm^3 cm^{-3} in the deep layer. This vertical distribution also confirmed the retention role of roots in the early growing season. The surface SW content increased significantly after August due to the precipitation and then was transported to the deep soil layers. Later, the impact of the crop roots decreased and the soil texture was the main factor for SW transport in September.

In order to further analyse SW content fluctuation in the temporal-vertical dimensions, a statistical analysis was performed on the data from the soil layers during the growing season (30 May–30 September) (Table 2). The SW content averages at the 15 cm and 60 cm depths were very close with values of 0.245 and 0.230 $cm^3\ cm^{-3}$, respectively. The mean SW content at the 30 cm and 90 cm depths were higher than at the other two layers with values of 0.272 $cm^3\ cm^{-3}$ and 0.280 $cm^3\ cm^{-3}$, respectively. The close relationship between the SW acquisition rate and the SW content that was shown in Fig. 2, was also confirmed by the statistical results. The SW acquisition rate had the same vertical pattern as the SW content and the higher values appeared at the 30 and 90 cm depths. Before maize seeding (May 30), the SW could be collected when the SW content reached about 0.25 $cm^3\ cm^{-3}$. After maize seeding, the SW could be collected when it reached a content of over 0.30 $cm^3\ cm^{-3}$. The SW acquisition rate was greatly and primarily influenced by rainfall.

Table 2: Statistical analysis of soil water content and acquisition rate at four depths during maize growing period

Depth (cm)								
Index	Soil water content ($cm^3\ cm^{-3}$)				Soil water acquisition rate ($cm^3\ s^{-1}$)			
	15	30	60	90	15	30	60	90
Minimum	0.210	0.235	0.205	0.240	0.00008	0.00005	0.00009	0.00003
Maximum	0.308	0.333	0.306	0.342	0.00241	0.00208	0.00324	0.00856
Mean	0.245	0.272	0.230	0.280	0.00125	0.00114	0.00140	0.00167
Std. Error	0.018	0.020	0.024	0.024	0.00077	0.00078	0.00091	0.00220
Coefficient of variance	0.074	0.072	0.106	0.088	0.615	0.685	0.648	1.320

Temporal Variation of Soil Water Storage

Based on the monitoring of the SW characteristics, the soil volumetric water content (SVW) and soil water storage (SWS) in the top 30 and

90 cm were calculated, respectively (Fig. 3). The SVW and SWS of the 0–90 cm and 0–30 cm depth showed similar temporal trends. In the vertical dimension, the SVW of the same two depths (0–30 cm and 0–90 cm) had similar trends in most periods, except in the first month which revealed a difference due to thawing. The thawing process from the surface to the deep layer and the tillage in preparation for sowing in the top soil decreased the SVW at 0–30 cm depth. There was almost no difference in the SVW between the two depths from mid-May to August. Before the crop harvest period in September, the deeper soil had a higher SVW content than the surface layer due to less root activity and strong precipitation. The SWS of the 0–30 cm depth had a seasonal variation and had a more direct relationship with precipitation events. Before the rainfall on 1 July, the SWS remained at a stable level from mid-May to the end of June. The SWS had a periodic decrease when no rainfall occurred in late July. The SWS reached its summit with the large rainfall in the later of August.

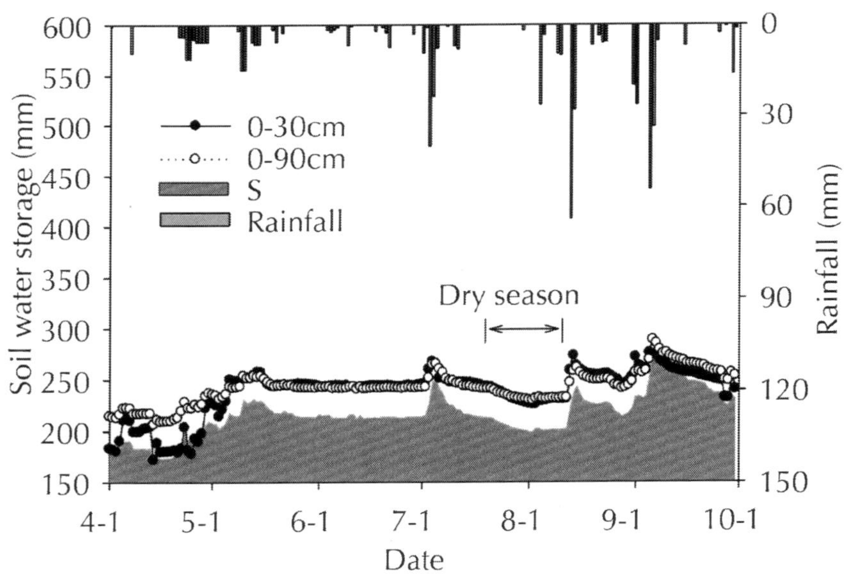

Figure 3: Temporal patterns of daily rainfall, soil water storage (SWS) and volumetric water content at 30 and 90 cm depths.

The SWS was calculated with the vertical variation in the SVW. The statistical characteristics of SWS and their proportions at each depth are listed in Table 3. The soil water storage averages at 0–20,

20–40 and 40–60 cm were 490.2, 593.8, and 358 m^3 ha^{-1}. The vertical differences indicated the middle level has higher soil water content as a consequence of soil texture and crop root zone influences. Because precipitation was the only water source for dryland, the SWS at the three layers in the dry period was smaller than the average for the entire growing season, which also revealed the impact of maize ET on SWS. The SWS in the vertical dimension had the same trend in the entire period and the middle layer had the largest value (548.9 m^3 ha^{-1}). The differences in SWS between the two periods become greater when going from the surface to the deeper layer, an observation derived from the rainfall contribution to the SWS in deeper layer.

Table 3: Dryland soil water storage and its proportion at three depths

	Duration		
Depth/ cm	Item	Entire growth season (30 May.-30 Sep.)	Dry season (20 Jul.-12 Aug.)
0-20	SWS/m^3 ha^{-1}	490.2±34.6	441.1 ±19.5
	Proportion/%	23.4± 1.04	23.3±0.4
0-40	SWS/m^3 ha^{-1}	1084±62.9	990.0±55.9
	Proportion/%	51.8±2.6	52.0±0.76
0-60	SWS/m^3 ha^{-1}	1442±111	1303.0±70.6
	Proportion/%	68.8±1.6	68.5±0.55

Based on daily monitoring data, the accumulative patterns of evapotranspiration (ET), rainfall, and water loss in the growing season were compared (Fig. 4). The ET was at a low level in the first month of the growing period and then assumed a stable increasing rate beyond the 50th day after sowing. The accumulated SW loss was consistent with the fluctuation in the rainfall. The ET and rainfall stayed with the SW loss over the first 22 days. The accumulated rainfall amount increased by 79.8 mm; a large increase occurred on the 35th day when the ET and water loss increased slightly. The rainfall increased intensively after 80 days, which also caused an increase in the SW loss. The accumulated amounts of rainfall increased from 269.3 mm after 80 days to 440.9 mm after 120 days, and the SW loss increased from 47.9 mm after 75 days to 126.3 mm after 120 days. The temporal pattern demonstrated that the SW loss and ET of dryland were strongly correlated with the rainfall. The ET was also a maize growth feature and its similar pattern with rainfall also proved it was the dominant factor for crop growth.

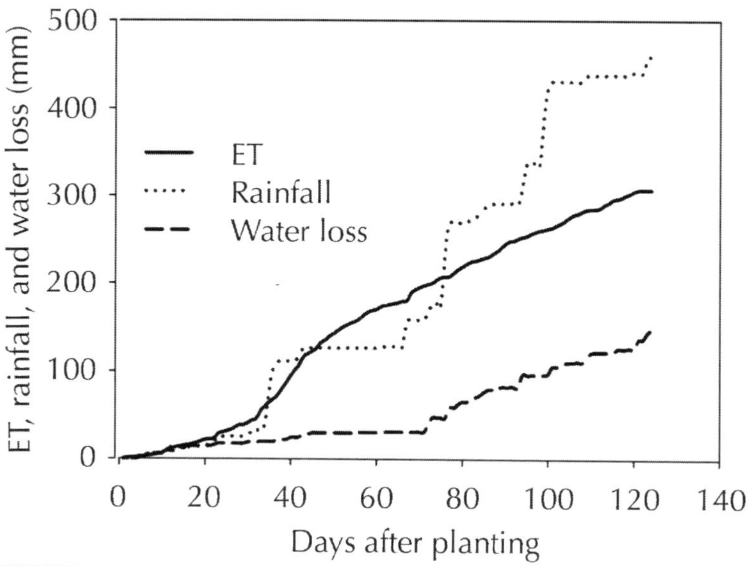

Figure 4: Cumulative evapo-transpiration (ET), rainfall, and water loss throughout maize growth period.

Variation of Crop Biomass in the Growing Period

The dry weight of the different maize organs (root, leaf, stem, tassel, and grain) in growing stages were analyzed (Fig. 5). Also shown in the Fig. 5 are the percentages of the total dry weight comprised by each organ in each of four growing seasons. The dry organ weight varied greatly from the seedling stage to maturity, a development period which is relatively short in this freeze-thawing area. The temporal pattern of dry weight in 2011 revealed the changes that occurred over four growing sections. The dry weight of grains increased significantly in the grain filling and maturity sections. The dry weight percentage did not change as intensively as the weight, and it was also found that the grains comprised nearly half of the maize weight in the maturity period.

With the N concentration in dried organs, the N absorption amount was quantified along with the organ dry weight. Due to logistical limitations, the crop organ sample in 2010 was only collected in the maturity period. The grain dry weights were larger in 2010 than in 2011. Comparing the dry weights of the organs in the maturity stage of the two years, the dry weights of the roots and tassel were significantly but their percentages differed little.

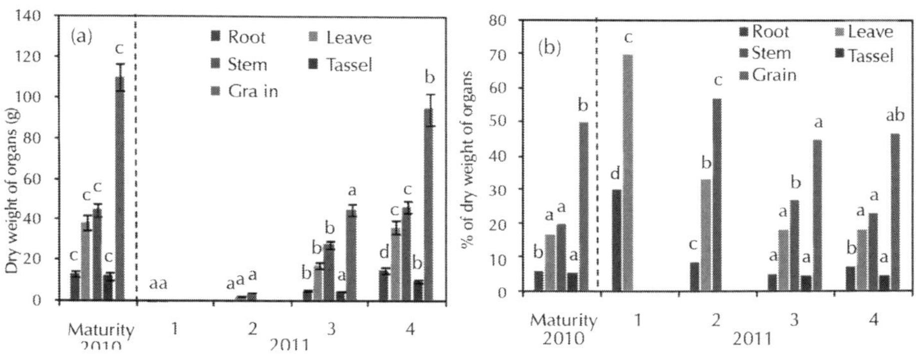

Figure 5: Dry weights (a) and percentage of maize organ weights (b) at different growth stages (Same letters in same organ between different growth stages indicate no significant difference.

1-Seedling, 2-Jointing, 3-Grain filling, 4-Maturity).

Nitrogen Uptake by Crop and Accumulation

The temporal N dynamics of crop uptake was then analyzed with the samples obtained from the maize organs (Fig. 6). As the maize grew, the N uptake in leaf first increased and then decreased with the maximum occurring in the jointing period. From grain filling to maturity, N uptake by leaf diminished insignificantly with a relative flat trend compared with that from the jointing to grain filling stages. As maize grew N uptake in stem decreased significantly decrease from 1 to 2 with a probable transfer of much N to tassel. The difference of N uptake by stem in the grain filling and maturity stages was insignificant, but the N uptake by grain increased significantly ($P<0.001$). The N uptake by leaf decreased insignificantly, which indicated that a large amount of N was transferred to tassel in the early time period and then transferred from tassel to grain (N uptakes by tassel in grain filling and maturity were 8.12 g kg^{-1} and 3.57 g kg^{-1}, respectively). There were insignificant differences in N uptakes of stem and leaf during the maturity stage in 2010 and 2011. The N uptake by grain in 2010 was 9.1% more than that in 2011.

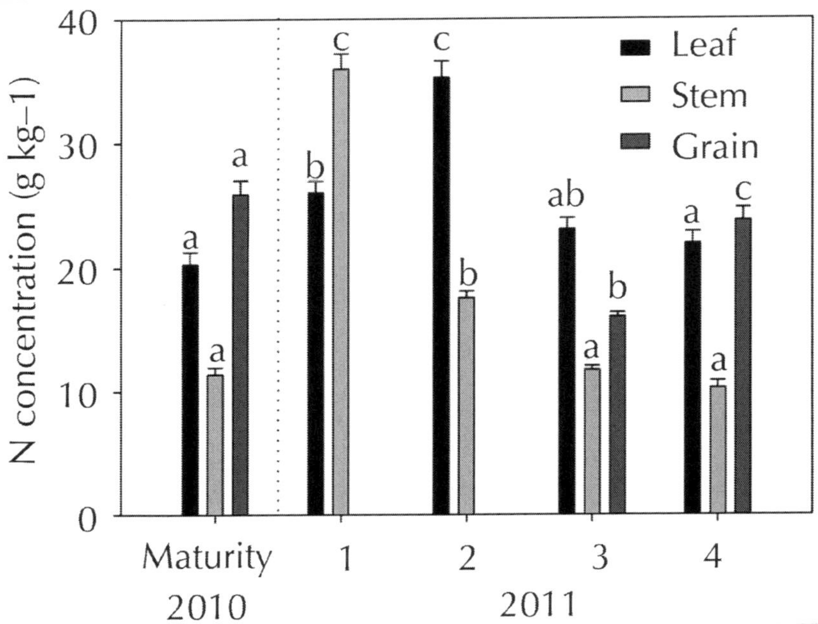

Figure 6: Seasonal dynamics of N concentration in maize organs (Same letters in same crop organs indicates no significant change between different growth stages.

1-Seedling, 2-Jointing, 3-Grain filling, 4-Maturity).

The accumulated N in different organs gradually increased in the early stages and reached a summit in the maturity period. From the jointing to grain filling period, the N in stem increased about 4.23 times and the amount in leaf increased less than five times. The N amounts in crop organs were not only related to the N uptake efficiency, but they also depended on the organ dry weight. According to the analysis in the maturity period, grain accumulated most of N, which occurred in a short time section. The total N amount of maize organs in 2010 was 309.9 kg ha^{-1}, which was larger than in 2011; the difference in grain was the dominant factor. The differences of N accumulation in leaf and stem during the maturity stages in the years 2010 and 2011 were insignificant.

Seasonal dynamics of N accumulative rates in organs during the growth period were analyzed (Fig. 7). From June to August, the N accumulative rates of stem and leaf increased rapidly and then declined. When the stem and leaf accumulated less N, the absorption efficiency in grain increased to 383 mg $m^{-2}d^{-1}$. The leaf was the most effective organ in uptaking N and the accumulation rate was about 500 mg $m^{-2}d^{-1}$.

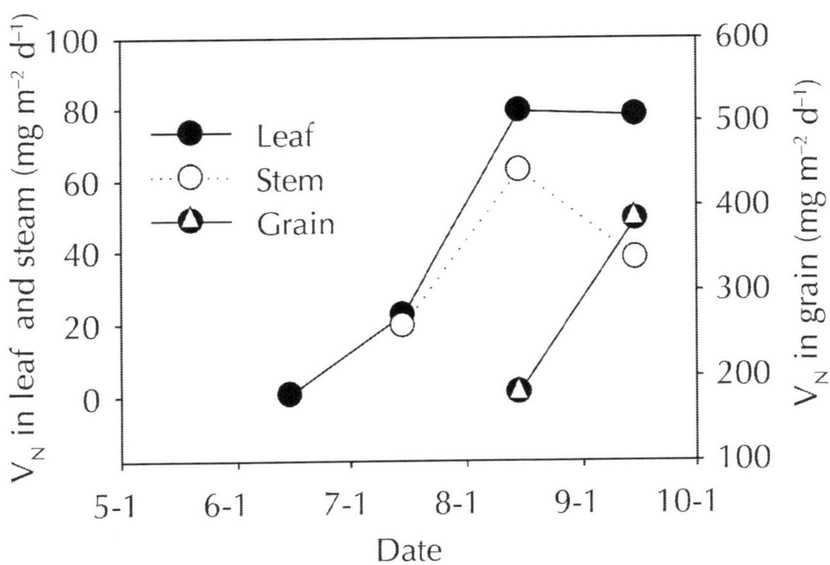

Figure 7: Seasonal dynamics of N accumulative rates (VN) in maize organs.

Variations of Soil N Stocks in the Growing Season

The N concentrations at four depths before sowing and after harvest were analyzed, which demonstrated the temporal-vertical soil N dynamics. The N (nitrate and ammonia) concentrations at different depths were first analyzed after the growing season (Table 5). Before the sowing, the ammonia-N stock at four depths ranged from 18.8 to 21.3 kg N $ha^{-1}$with no occurrence of a significant difference. After the harvest, the deeper layer experienced more loss than the surface layer. The middle two layers had close values and the 12.5 kg N ha^{-1} in top layer was the largest value. The nitrate-N concentration in soil was much larger than the ammonia-N. At the beginning of the growing season, the nitrate-N concentration was more than 121 kg ha^{-1} at all depths and had slight differences at 60–90 cm depth. The nitrate-N concentration decreased to 90 kg ha^{-1} after the harvest. The upper two layers had similar patterns and the concentrations ranged between 88–90 kg ha^{-1}. There was no significant difference in the concentration in the deeper two layers and their concentrations ranged from 63.2 to 68.0 kg ha^{-1}. The nitrate-N was more variable than ammonia-N and had a larger stock gap.

Table 4: Temporal pattern of N amount (kg ha^{-1}) of maize organs during growing season*

Year	Growth stage	Leaf	Stem	Grain
2010	Mature	58.6f	38.3ef	213de
2011	Seedling	0.14a		
	Jointing	6.85c	5.81a	
	Grain filling	30.6d	24.6d	54.1b
	Maturity	60.0f	36.0e	169c

Table 5: Mineral N stocks (kg ha^{-1}) and variations at each depth of growing season

Depth (cm)	Before sowing		After harvest		Stock gap	
	NO_3^--N	NH_4^+-N	NO_3^--N	NH_4^+-N	NO_3^--N	NH_4^+-N
0-15	147d	21.3b	90.1d	12.5c	-48.9c	-8.64bc
15-30	144d	21.2b	88.1d	8.12b	-53.8cd	-11.2c
30-60	130cd	19.3b	68.0c	9.02b	-65.3e	-9.79c
60-90	121c	18.8b	63.2c	5.74a	-56.8d	-12.69c

Nutrient Loss and Supply in Soil-Crop System

Based on field water monitoring and SW content in the 0–90 cm profile, the soil water loss (SL) and the leached N were calculated (Table 6). The rainfall and evapo-transpiration (ET) were calculated in the four stages of the maize growing season. The SL had a closer correlation with rainfall than with ET. The rainfall was nearly 100 mm in the jointing stage, but the SL was 11.5 mm due to the intensive ET. Moving from the grain filling stage to the maturity stage, the ET increased and more SL occurred under the higher rainfall event. The leached N was also calculated with the N concentration in soil water. The nitrate-N was the dominant N lost from the 90 cm soil profile, especially in the grain filling and maturity stages. The leached ammonia-N had similar temporal patterns and the largest amount was 0.18 kg ha^{-1}.

Table 6: Amounts of soil water loss (SL) and N leached below maize root zone (90 cm)

Growth stage	Period	R	ET	SL	leached N (kg ha^{-1})	
		(mm)			NO_3^--N	NH_4^+-N
Seedling	30 May to 29 Jun.	29.8	44.6	19.6	0.03	0.01
Jointing	30 Jun. to 25 Jul.	96.5	119.0	11.5	3.36	0.12
Grain filling	26 Jul. to 23 Aug.	165	67.3	48.6	5.53	0.18
Maturity	24 Aug. to 30 Sep.	168	76.6	67.5	6.26	0.10

Based on the N monitoring of rainfall and the survey of N fertilization, the N in rainfall and fertilizer was calculated. With the SWS and N concentration in SW in 0–90 cm profile, the *NUE*in the maize-soil system was calculated (Table 7). The N input from rainfall and fertilizer in the entire maize growing season was 16.5 and 120 kg ha^{-1}, respectively. The soil N supply in the entire growth period was about 641 kg ha^{-1}. With the nitrate and ammonia concentrations in SW and water volume, the N loss was estimated at 15.6 kg ha^{-1}. Based on the N balance, it was calculated that the *NUE* was 43%. On the other hand, the excessive N from dryland was an important potential source for N discharge to the environment.

Table 7: The N balance and usage efficiency (*NUE*) in crop-soil system during the maize growth period

Rainfall (kg N ha^{-1})	Fertilizer (kg N ha^{-1})	Initial soil N (kg N ha^{-1})		Final soil N (kg N ha^{-1})		N loss (kg N ha^{-1})		Soil supply N (kg N ha^{-1})	NUE (kg kg^{-1})
		NO_3^--N	NH_4^+-N	NO_3^--N	NH_4^+-N	NO_3^--N	NH_4^+-N		
16.5	120	542.9 ± 29.76	80.6 ± 4.29	309.4 ± 18.4	35.4 ± 4.37	15.2	0.41	641 ± 14.4	0.43 ± 0.03

DISCUSSION

The Dryland Soil Hydrological Process in a Freeze-Thawing Area

Rainfall is the only water source in dryland and it has a direct impact on SW content and storage [33]. The field monitoring done in four stages as part of our study indicated that the temporal patterns of SW content and acquisition rate closely followed the rainfall amount (Fig. 2). The soil roughness in a cultivated area is increased by tillage, which enhances the vertical transport of SW [34]. The SL was 146.7 mm (32% of rainfall in the growing season). However, the vertical distribution of SW content and soil water storage demonstrate that the water has a short vertical movement range. The SW can be sampled only after the rainfall events, but the acquisition rates showed slight vertical differences. These two findings both demonstrated that the vertical movement of water is slow even with large volumes and the speed can affect the range.

Compared with the long growing periods in other dryland agricultural areas, the local maize growing season is short and has a higher efficiency in the use of water and N [35]. A proper SW condition in the root zone is essential for crop growth in the maize seeding and grain filling stages. The SWS can remain at a stable level (about 1000 m^3 ha^{-1} of 40 cm profile) even in the dry season without precipitation (Fig. 3). All of these factors indicate that the *albic* soil texture can provide a positive water condition for crop growth [24]. The SW also controls surficial N movements in the dryland and provides a strong signal for crop absorption [36]. Crop transpiration will increase when the maize is near its maturity period, and the precipitation normally increases and meets this need during the same period (Fig. 4). The combined conditions of special soil texture and precipitation patterns directly determine the dryland soil hydrological pattern, which is the critical factor for maize growth in this freeze-thawing area.

Crop Growth and N Accumulation Differences in Plant Organs

The quantification and assessment of the organ biomass in crops can help in understanding the impacts of SW and N on crop growth. This information is also necessary for assessing N leaching [37]. The crop growth is measured by the organ dry weight (Fig. 5), which indicates the crop growth rate is higher than in other drylands under warming temperatures. The yearly precipitation in 2011 was more than in 2000 (Fig. 2), but did not lead to the higher grain dry weight in 2011 (Fig. 5). The difference proves that a higher maize yield is based on a combination of climate, soil and tillage. The grain filling period in this freeze-thawing area occurs in August, at a time when temperature and precipitation are at their highest levels. The active ET facilitates the transport of water and N from soil to crop, which finally causes a higher grain filling performance [38].

The absorbed N by maize organs in the entire growth period had different allocation rates (Fig. 7), which is determined by the photosynthesis of the organs and the eco-hydrological process. The crop accumulative rate in its canopy, stem and grain varied significantly, which means the N absorption capability for tissue production of dedicated organs shifts due to the N flux from soil to above ground biomass [39]. The soil N concentration in the vertical direction proved that the subsurface has a significant variance due to the root uptake. The accumulation rates of different organs over time also follow the typical sigmoid curve [40]. The temporal patterns of N in maize organs also indicate the N absorption is not only related to the soil N concentration, but that there is a complicated systemic feedback signal within the soil-crop system [41]. The SW availability, ET and N concentration are the key signaling factors. As the soil N is transported in crops with water, there are plenty of field observations and models to identify the growth stress reaction to the SW and N conditions [42],[43]. This information provides detailed guidelines for N and water adjustment in the soil-crop system.

Implication for N Leach Control and Soil N Optimization

The N uptake efficiency of a crop is the main component for *NUE* and the temporal patterns inTable 4 indicated that the highest amount of grain uptake of N occurs in the graining filling and maturity periods. The maize uptake of N will increase until the optimum levels of water and N in soils are reached [44]. Overlaying this information with the leached N loading, it was found that the intensive N uptake and loss occurred at same time. The strong ET and rainfall in that time period were the reason for the N transport in the crop-soil system [45]. The soil N gap analysis revealed that the observed N gap was similar in the vertical direction, which also demonstrates that the soil N can move among the soil depths with hydrological processes and maintain the N concentration balances [3]. The temporal N analysis in crop organs, soil and leached water explain the maize growth in this freeze-thawing area, which also provides a guideline for N fertilization while, also considering SW conditions.

The N pollution to groundwater due to excessive N fertilizer application is an important environmental concern [46]. The yearly *NUE* is about 43% in this area, which is a relatively high value in comparison with similar dryland agriculture areas [47]. The N leached from farmland mainly depends on the difference between the N obtained by the soil-water-crop system and the N uptake by the crop

[48]. As Table 6 shows, the leached N is mainly related to the loss of SW, which is the gap between rainfall and ET. The rainfall in August 2011 was above the long term monthly value, which caused the highest SW and N loss during the growing season. Under the normal climatic conditions, the *NUE* in this area can reach 50%. The nitrate-N is the dominant form of leached N to the groundwater, which is a direct consequence of fertilization and soil nitrification [49]. Models have been developed to optimize the fertilizer application to farmland based on a consideration of the soil hydrological process, which can help to maximize the crop yield with minimal discharge to the environment [50]. The model with local characteristic parameters, should be applied in the tillage practice to obtain an improved *NUE*.

CONCLUSION

The results of this study have shown that combined conditions of soil texture, soil hydrology and precipitation were basic factors for maize growth in the freeze-thawing area in Northeast China. The rainfall significantly affected the pattern of SW content and the allocation of N in the maize organs. Precipitation was the only water source for the dryland, but the variation in SW acquisition rates and SW content remained at a stable level due to the special *albic soil* texture. The dry weight and N accumulation of maize organs in the entire growing period proved that the soil hydrological condition was also enhanced.

The N loss was dependent on both the hydrological process and the N in the water, which was then decided by the rainfall water depth and the applied N. The NO_3-N was the dominant leachate, which was closely correlated with the amount of applied N fertilizer. The *NUE* was about 43%, which was also related to the initial soil N and maize uptake. In order to improve the*NUE* and decrease the amounts of leached nitrogen, it would be better to improve the fertilizer application amounts while giving consideration to rainfall patterns and N accumulation rates.

REFERENCES

1. Buckley C, Carney P (2013) The potential to reduce the risk of diffuse pollution from agriculture while improving economic performance at farm level. Environ Sci Policy 25(1): 118–126. doi: 10.1016/j.envsci.2012.10.002
2. Quemada M, Baranski M, Nobel-de Lange MNJ, Vallejo A, Cooper JM (2013) Meta-analysis of strategies to control nitrate leaching in irrigated agricultural systems and their effects on crop yield. Agric Ecosyst Environ 174(15): 1–10. doi: 10.1016/j.agee.2013.04.018
3. Ferrant S, Oehler F, Durand P, Ruiz L, Salmon-Monviola J, et al. (2011) Understanding nitrogen transfer dynamics in a small agricultural catchment: Comparison of a distributed (TNT2) and a semi distributed (SWAT) modeling approaches. J Hydrol 406(1–2): 1–15. doi: 10.1016/j.jhydrol.2011.05.026
4. Panda RK, Behera S (2003) Non-point source pollution of water resources: Problems and perspectives. J Food Agric Environ 1(3–4): 308–311.
5. Wang XB, Dai KA, Wang Y, Zhang XM, Zhao QS, et al. (2010) Nutrient management adaptation for dryland maize yields and water use efficiency to long-term rainfall variability in China. Agric Water Manage 97(9): 1344–1350. doi: 10.1016/j.agwat.2010.03.016
6. Yang YG, Xiao HL, Wei YP, Zhao LJ, Zou SB, et al. (2011) Hydrologic processes in the different landscape zones of Mafengou River basin in the alpine cold region during the melting period. J Hydrol 409(1–2): 149–156. doi: 10.1016/j.jhydrol.2011.08.013
7. Grzebisz W (2013) Crop response to magnesium fertilization as affected by nitrogen supply. Plant Soil 368(1–2): 23–39. doi: 10.1007/s11104-012-1574-z
8. Matejek B, Huber C, Dannenmann M, Kohlpaintner M, Gasche R, et al. (2010) Microbial N turnover processes in three forest soil layers following clear cutting of an N saturated mature spruce stand. Plant Soil 337(1–2): 93–110. doi: 10.1007/s11104-010-0503-2

9. Moreau P, Ruiz L, Vertes F, Baratte C, Delaby L, et al. (2013) CASIMOD'N: An agro-hydrological distributed model of catchment-scale nitrogen dynamics integrating farming system decisions. Agric Syst 118: 41–51. doi: 10.1016/j.agsy.2013.02.007
10. Kersebaum KC, Steidl J, Bauer O, Piorr HP (2003) Modelling scenarios to assess the effects of different agricultural management and land use options to reduce diffuse nitrogen pollution into the river Elbe. Phys Chem Earth 28: 537–545. doi: 10.1016/s1474-7065(03)00090-1
11. Hirel B, Le Gouis J, Ney B, Gallais A (2007) The challenge of improving nitrogen use efficiency in crop plants: towards a more central role for genetic variability and quantitative genetics within integrated approaches. J Exp Bot 58: 2369–2387. doi: 10.1093/jxb/erm097
12. Pask AJD, Sylvester-Bradley R, Jamieson PD, Foulkes MJ (2012) Quantifying how winter wheat crops accumulate and use nitrogen reserves during growth. Field Crop Res 126(14): 104–118. doi: 10.1016/j.fcr.2011.09.021
13. Bonato O, Schulthess F, Baumgärtner J (1999) Simulation model for maize crop growth based on acquisition and allocation processes for carbohydrate and nitrogen. Ecol Modell 124(1): 11–28. doi: 10.1016/s0304-3800(99)00146-5
14. Binder DL, Dobermann A, Sander DH, Cassman KG (2002) Biosolids as nitrogen source for irrigated maize and rainfed sorghum. Soil Sci Soc Am J 66(2): 531–543. doi: 10.2136/sssaj2002.0531
15. Favaretto N, Norton LD, Johnston CT, Bigham J, Sperrin M (2012) Nitrogen and Phosphorus Leaching as Affected by Gypsum Amendment and Exchangeable Calcium and Magnesium. Soil Sci Soc Am J 76(2): 575–585. doi: 10.2136/sssaj2011.0223
16. Lam QD, Schmalz B, Fohrer N (2012) Assessing the spatial and temporal variations of water quality in lowland areas, Northern Germany. J Hydrol 438–439(17): 137–147. doi: 10.1016/j.jhydrol.2012.03.011
17. Friend AD, Stevens AK, Knox RG, Cannell MGR (1997) A process-based, terrestrial biosphere model of ecosystem dynamics

(Hybrid v3.0). Ecol Modell 95(2–3): 249–287. doi: 10.1016/s0304-3800(96)00034-8

18. Dorioz JM, Ferhi A (1994) Non-point pollution and management of agricultural areas: Phosphorus and nitrogen transfer in an agricultural watershed. Water Res 28(2): 395–410.
19. Keeney DR, Follett RF (1991) Managing Nitrogen for Groundwater Quality and Farm Profitability, SSSA, Madison, WI. 1–7.
20. Kim DG, Vargas R, Bond-Lamberty B, Turetsky MR (2012) Effects of soil rewetting and thawing on soil gas fluxes: a review of current literature and suggestions for future research. Biogeoscience 9(7): 2459–2483. doi: 10.5194/bg-9-2459-2012
21. Lewis DR, McGechan MB, McTaggart IP (2003) Simulating field-scale nitrogen management scenarios involving fertiliser and slurry applications. Agr Syst 76(1): 159–180. doi: 10.1016/s0308-521x(02)00032-x
22. Dawson JC, Huggins DR, Jones SS (2008) Characterizing nitrogen use efficiency in natural and agricultural ecosystems to improve the performance of cereal crops in low-input and organic agricultural systems. Field Crop Res 107: 89–101. doi: 10.1016/j.fcr.2008.01.001
23. Schroder JL, Zhang HL, Girma K, Raun WR, Penn CJ, et al. (2011) Soil Acidification from Long-Term Use of Nitrogen Fertilizers on Winter Wheat. Soil Sci Soc Am J 75(3): 957–964. doi: 10.2136/sssaj2010.0187
24. Caliskan S, Ozkaya I, Caliskan ME, Arslan M (2008) The effects of nitrogen and iron fertilization on growth, yield and fertilizer use efficiency of soybean in a Mediterranean-type soil. Field Crop Res 108(2): 126–132. doi: 10.1016/j.fcr.2008.04.005
25. Razzaghi F, Plauborg F, Jacobsen SE, Jensen CR, Andersen MN (2012) Effect of nitrogen and water availability of three soil types on yield, radiation use efficiency and evapotranspiration in field-grown quinoa. Agric Water Manage 109: 20–29. doi: 10.1016/j.agwat.2012.02.002
26. Hao FH, Chen SY, Ouyang W (2013) Temporal rainfall patterns with water partitioning impacts on maize yield in a freeze-thaw zone. J Hydrol 486(12): 412–419. doi: 10.1016/j.jhydrol.2013.02.008

27. Ouyang W, Huang HB, Hao FH, Shan YS, Guo BB (2012) Evaluating spatial interaction of soil property with non point source pollution at watershed scale: the phosphorus indicator in Northeast China. Sci Total Environ 432: 412–421. doi: 10.1016/j.scitotenv.2012.06.017
28. Jackson JE (1991) A user's guide to principal components. Wiley-Interscience, New York.
29. Moroizumi T, Hamada H, Sukchan S, Ikemoto M (2009) Soil water content and water balance in rainfed fields in Northeast Thailand. Agric Water Manage 96(1): 160–166. doi: 10.1016/j.agwat.2008.07.007
30. Huggins DR, Pan L (1993) Nitrogen efficiency component analysis: an evaluation of cropping system differences in productivity. Agronomy J 85: 898–905. doi: 10.2134/agronj1993.00021962008500040022x
31. Meisinger JJ, Randall GW (1991) Estimating nitrogen budgets for soil-crop systems. In: Follet, R.F., et al. (Eds.), Managing Nitrogen for Groundwater Quality and Farm Profitability. SSSA, Madison,WI, USA, 85–124.
32. Vázquez N, Pardo A, Suso ML, Quemada M (2006) Drainage and nitrate leaching under processing tomato growth with drip irrigation and plastic mulching. Agric Ecosyst Environ 112(4): 313–323. doi: 10.1016/j.agee.2005.07.009
33. Tanveer SK, Wen XX, Lu XL, Zhang JL, Liao YC (2013) Tillage, Mulch and N Fertilizer Affect Emissions of CO2 under the Rain Fed Condition. PLOS ONE 8(9): e72140. doi: 10.1371/journal.pone.0072140
34. Dörner J, Horn R (2009) Direction-dependent behaviour of hydraulic and mechanical properties in structured soils under conventional and conservation tillage. Soil Till Res 102(2): 225–232. doi: 10.1016/j.still.2008.07.004
35. Murungu FS, Chiduza C, Muchaonyerwa P, Mnkeni PNS (2011) Mulch effects on soil moisture and nitrogen, weed growth and irrigated maize productivity in a warm-temperate climate of South Africa. Soil Till Res 112(1): 58–65. doi: 10.1016/j.still.2010.11.005

36. Gheysari M, Mirlatifi SM, Homaee M, Asadi ME, Hoogenboom G (2009) Nitrate leaching in a silage maize field under different irrigation and nitrogen fertilizer rates. Agric Water Manage 96(6): 946–954. doi: 10.1016/j.agwat.2009.01.005
37. Wegehenkel M, Mirschel W (2006) Crop growth, soil water and nitrogen balance simulation on three experimental field plots using the Opus model—A case study. Ecol Modell 190(1-2): 116–132. doi: 10.1016/j.ecolmodel.2005.02.020
38. Katerji N, Campi P, Mastrorilli M (2013) Productivity, evapotranspiration, and water use efficiency of corn and tomato crops simulated by AquaCrop under contrasting water stress conditions in the Mediterranean region. Agric Water Manage 130: 14–26. doi: 10.1016/j.agwat.2013.08.005
39. Strullu L, Cadoux S, Preudhomme M, Jeuffroy MH, Beaudoin N (2011) Biomass production and nitrogen accumulation and remobilisation by Miscanthus × giganteus as influenced by nitrogen stocks in belowground organs. Field Crop Res 121(3): 381–391. doi: 10.1016/j.fcr.2011.01.005
40. Shoji S, Gandeza AT, Kimura K (1991) Simulation of crop response to polyolefin-coated urea: II. Nitrogen uptake by corn. Soil Sci Soc Am J 55(5): 1468–1473. doi: 10.2136/sssaj1991.03615995005500050045x
41. Nacry P, Bouguyon E, Gojon A (2013) Nitrogen acquisition by roots: physiological and developmental mechanisms ensuring plant adaptation to a fluctuating resource. Plant Soil 370(1–2): 1–29. doi: 10.1007/s11104-013-1645-9
42. Kovács GJ (2005) Modelling of adaptation processes of crops to water and nitrogen stress. Phys Chem Earth, Parts A/B/C 30(1–3): 209–216. doi: 10.1016/j.pce.2004.08.022
43. Overman AR, Scholtz RV (2013) Accumulation of Biomass and Mineral Elements with Calendar Time by Cotton: Application of the Expanded Growth Model. PLOS ONE 8(9): e72810. doi: 10.1371/journal.pone.0072810
44. Kerbiriou PJ, Stomph TJ, Van Der Putten PEL, Van Bueren ETL, Struik PC (2013) Shoot growth, root growth and resource capture under limiting water and N supply for two cultivars of lettuce (Lactuca sativa L.). Plant Soil 371(1–2): 281–297. doi: 10.1007/s11104-013-1672-6

45. Asseng S, Turner NC, Keating BA (2001) Analysis of water- and nitrogen-use efficiency of wheat in a Mediterranean climate. Plant Soil 233(1): 127–143. doi: 10.1023/a:1010381602223
46. Fang QX, Yu Q, Wang EL, Chen YH, Zhang GL, et al. (2006) Soil nitrate accumulation, leaching and crop nitrogen use as influenced by fertilization and irrigation in an intensive wheat-maize double cropping system in the North China Plain. Plant Soil 284(1–2): 335–350. doi: 10.1007/s11104-006-0055-7
47. Ouyang W, Wei XF, Hao FH (2013) Long-term soil nutrient dynamics comparison under smallholding land and farmland policy in northeast of China. Sci Total Environ 450–451(15): 129–139. doi: 10.1016/j.scitotenv.2013.02.016
48. Hall JA, Bobe G, Hunter JK, Vorachek WR, Stewart WC, et al. (2013) Effect of Feeding Selenium-Fertilized Alfalfa Hay on Performance of Weaned Beef Calves. PLOS ONE 8(3): e58188. doi: 10.1371/journal.pone.0058188
49. Lteif A, Whalen JK, Bradley RL, Camire C (2010) Nitrogen transformations revealed by isotope dilution in an organically fertilized hybrid poplar plantation. Plant Soil 333(1–2): 105–116. doi: 10.1007/s11104-010-0326-1
50. Sierra J, Brisson N, Ripoche D, Noel C (2003) Application of the STICS crop model to predict nitrogen availability and nitrate transport in a tropical acid soil cropped with maize. Plant Soil 256(2): 333–345.

Chapter 5

Two Challenges for U.S. Irrigation Due to Climate Change: Increasing Irrigated Area in Wet States and Increasing Irrigation Rates in Dry States

Robert I. McDonald[1*], Evan H. Girvetz[2]

[1]Central Science Program, The Nature Conservancy, Arlington, Virginia, United States of America,

[2]Global Climate Change Program, The Nature Conservancy, Seattle, Washington, United States of America

ABSTRACT

Agricultural irrigation practices will likely be affected by climate change. In this paper, we use a statistical model relating observed water use by U.S. producers to the moisture deficit, and then use this statistical model to project climate changes impact on both the fraction of agricultural land irrigated and the irrigation rate (m^3ha^{-1}). Data on water withdrawals for US states (1985–2005) show that both quantities are highly positively correlated with moisture deficit (precipitation – PET). If current trends hold, climate change would increase agricultural demand for irrigation in 2090 by 4.5–21.9 million ha (B1 scenario demand: 4.5–8.7 million ha, A2 scenario demand: 9.1–21.9 million ha). Much of this new irrigated area would occur in states that currently have a wet climate and a small fraction of their agricultural land currently irrigated, posing a challenge to policymakers in states with less experience with strict regulation of agriculture water use. Moreover, most of this expansion will occur in states where current agricultural production has relatively low market value per hectare, which may make installation of irrigation uneconomical without significant changes in crops or practices by producers. Without significant increases in irrigation efficiency, climate change would also increase the average irrigation rate from 7,963 to 8,400–10,415 m^3ha^{-1} (B1 rate: 8,400–9,145 m^3ha^{-1}, A2 rate: 9,380–10,415 m^3ha^{-1}). The irrigation rate will increase the most in states that already have dry climates and large irrigation rates, posing a challenge for water supply systems in these states. Accounting for both the increase in irrigated area and irrigation rate, total withdrawals might increase by 47.7–283.4 billion m^3 (B1 withdrawal: 47.7–106.0 billion m^3, A2 withdrawal: 117.4–283.4 billion m^3). Increases in irrigation water-use efficiency, particularly by reducing the prevalence of surface irrigation, could eliminate the increase in total irrigation withdrawals in many states.

INTRODUCTION

Anthropogenic emissions of greenhouse gases are very likely increasing average temperatures and are more likely than not altering the amount and timing of precipitation [1]. The effects of climate change on agriculture will likely be multifaceted [2]–[5]. Changes in temperature, precipitation [6] and CO_2 concentration effects will affect plant growth. Moreover, the indirect effects of changes in weed and pest abundance and disease outbreaks may also significantly affect agriculture [7], [8]. This paper focuses on irrigation, the largest human water use, which will likely be affected by climate change [9].

Climate change will likely affect the supply of water available from surface water or groundwater, decreasing it in many cases, and may also increase evapotranspiration from crops and hence demand for irrigation [10]–[16]. While several papers have quantified how changes in climate will affect irrigation water use by agricultural producers, results have been mixed depending on the time horizon of the analysis, the particular general circulation models (GCMs) consulted, and assumptions about how producers' irrigation practices respond to climate change [17]–[20].

The decision of whether or not to irrigate an agricultural field, as well as the amount of irrigation water applied, is a complex decision for producers, and depends on a number of factors [18],[21]. Water supply is constrained by water policy and law as well as the infrastructure available to bring water to farmers. The potential economic return from production depends on the crops grown, the productivity of the soil, and the other inputs farmers add such as fertilizer. Demand for irrigation also depends on the crop grown, as well as the prevailing climate. Climate change might increase moisture deficit, or decrease it, depending on the relative change in temperatures and precipitation. Accounting for all these factors is difficulty, which makes forecasting irrigation water use challenging.

This paper uses an empirical approach, showing for the United States that changes in climate and irrigation equipment over the past 25 years have resulted in predictable changes in both the fraction of agricultural land irrigated and the irrigation rate (m^3 ha^{-1}). We then use a simple statistical model to make state-level projections of how both quantities will be affected by climate change. We incorporate information on historical changes over time in irrigation equipment into our statistical model, and project the potential scope of future irrigation efficiency improvements to serve as an adaptation to climate change. We use this analysis to estimate the how irrigation rate and area irrigated will need to change in the future–based on the assumptions of this model–to keep up with the change in water demand of crops due to climate change.

MATERIALS AND METHODS

Data

Data on historical withdrawals for the period 1985–2005 were obtained from the United States Geological Survey (USGS), which conducts county-level surveys every five years [22]. For this study, we use total freshwater withdrawals (surface water plus groundwater) by the agriculture sector. For our analysis, we lump county-level data to states, since the county-level data vary widely in withdrawals when the county in which an irrigation system withdraws water is different than the county where the water is applied.

Withdrawal information was supplemented by data on area irrigated, as defined in the United States Department of Agriculture (USDA) Farm and Ranch Irrigation Survey [23], and total agricultural area and its market value, as defined in the USDA Census of Agriculture [24]. In this study, our definition of "agricultural area" follows the USDA definition of "total cropland", which includes harvested cropland, cropland used for pasture or grazing, and other miscellaneous cropland (e.g., areas where crops failed, fields were left intentionally left fallow, or with cover crops). Note that the subcategory of "cropland used for pasture and grazing" includes only land currently used for pasture or grazing that could be immediately used for crops without any additional improvement; it does not include the much larger land area of woodland and rangeland that is used for grazing or pasturing.

Data on historical temperature and precipitation were taken from the Parameter-elevation Regressions on Independent Slopes Model (PRISM) dataset (~4 km resolution) and used to calculate historic observed moisture deficit [25]. Future climate scenarios are based on the ensemble mean of a panel of 16 general circulation models (GCM) [26], [27] for each of the A2, A1B, and B1 scenarios of the Intergovernmental Panel on Climate Change [28]. The 16 GCMs used in this study are BCCR-BCM2.0, CGCM3.1(T47), CNRM-CM3, CSIRO-Mk3.0, GFDL-CM2.0, GFDL-CM2.1, GISS-ER, INM-CM3.0, IPSL-CM4, MIROC3.2(medres), ECHO-G, ECHAM5/MPI-OM, CCSM3, PCM, and UKMO-HadCM3. For more information on the structure of each GCM, see Meehl et al. [26]. We examined three time periods:

2020–2039 (hereafter "2030"), 2040–2059 (hereafter "2050"), 2060–2079 (hereafter "2070"), and 2080–2099 (hereafter "2090"). To capture uncertainty in the GCM predictions of climate under different emissions scenarios, we also include the 20% and 80% quintiles of the ensemble's prediction of moisture deficit.

By using an ensemble of 16 GCMs, and considering the variation among the ensemble predictions for GCMs, we hope to improve on previously published work by considering a broad set of GCMs. Other papers that have quantified how changes in climate will affect irrigation water use have only considered a few GCMs, and some of the variation in the results among and within the papers seems to be driven by the particular GCMs chosen [17]–[20]. For instance, Fisher et al. [17] looked at the climate change impacts of the a2r emissions scenario with the HadCM3 and CISRO GCMs, over the time period 1990–2080, projecting a 45% increase in irrigated area globally. Their predictions for North America varied between the two GCMs, because the CSIRO MK3 predicts a decrease in moisture deficit for much of the US, while the Hadley CM3 predicts an increase in moisture for much of the U.S. Similarly, Thomson et al. [19] looked at the climate change impacts of different average temperature increases with two GCMs, the HadCm3 and CGCM, but incorporated in their model both demand for irrigation and supply of available water. They predict that irrigated area will decline in the continental United States under both GCMs, despite strong growth in demand, because of a decline in the available supply of irrigation water.

Calculation of Moisture Deficit

We calculated the soil moisture deficit during the growing season by subtracting precipitation from potential evapotranspiration (PET), as measured with the modified Thornthwaite (Hamon) method [29], which is based on temperature and number of daylight hours, and widely used in global and regional hydrologic models [30]. This metric should theoretically be related to irrigation water required. Moreover, it integrates intra-annual variability in water demand into a cumulative number for the growing season.

GCM predictions of moisture deficit were first resampled to average county values, and then average state values were calculated, weighting by the agricultural area (for the analysis of fraction of area irrigated) or irrigated area (for the analysis of irrigated area).

There has been considerable debate about the merits of different methods of estimating PET, and Kingston et al. [31] showed that the relative magnitude of the effect of climate change on PET depends on which method is used. The Hamon method appears to be one of the more sensitive metrics, increasing with climate change greatly, whereas other methods like Blaney-Criddle and Priestley-Taylor increase less [31]. To ensure that our results were not influence by our choice of metrics, we also calculated PET using the Blaney-Criddle method, a simple method that is known to be relatively less sensitive to climate change than most metrics [31]. The Blaney-Criddle method is a function solely of the mean daily temperature and the mean daily percentage of daytime hours.

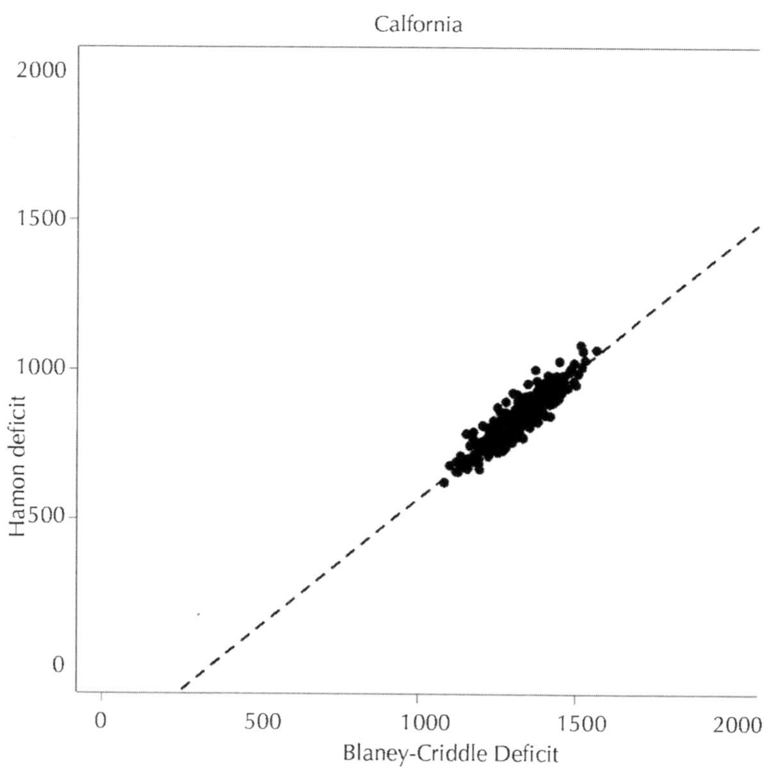

Figure 1: Scatterplot of estimated PET for California in 2005, using the Hamon and Blaney-Criddle metrics.

Preliminary results suggest no major change in our findings. To see why this is likely to be so, note that we build a statistical model that relates irrigation use by farmers to historical values of PET. Since Hamon and Blaney-Criddle give highly correlated estimates of PET (Figure 1), and changes in Hamon and changes in Blaney-Criddle are highly correlated (Figure 2), they both give similar results in the linear regression we use as our statistical model. While change in PET as estimated by Hamon is systematically higher than change in PET by Blaney-Criddle, the fitted slopes in our regression correct for this fact and given similar estimates of irrigation use. Because we are fitting either estimate of PET to empirical data on irrigation use, as long as the estimates are strongly linearly correlated with one another, the projections of the statistical model will be very similar.

Each dot represents one year in a particular combination of GCM and greenhouse gas emissions Scenario. Note that the strong linear correlation between the two means that when either of these two metrics are statistically related to irrigation use, the quantitative predictions of the effect of climate change are quite similar. Other states also show a linear correlation, with R^2 ranging from 0.75 to 0.93.

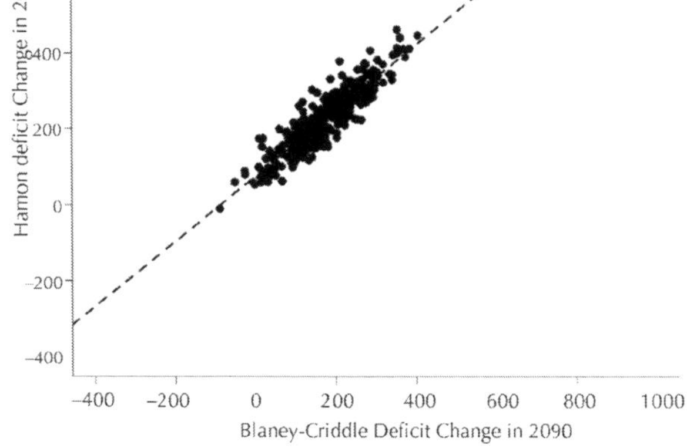

Figure 2: Scatterplot of change in PET for California between 2005 and 2090, using the Hamon and Blaney-Criddle metrics.

Each dot represents one year in a particular combination of GCM and greenhouse gas emissions scenario. Note that the strong linear correlation between the two means that when either of these two metrics are statistically related to irrigation use, the quantitative predictions of the effect of climate change are quite similar. Other states also show a linear correlation, with R^2 ranging from 0.74 to 0.93.

The spatial patterns of change in Hamon and Blaney-Criddle are quite similar (Figure 3). However, there are slight differences among states. For example, the Hamon metric predicts higher increases in PET in arid parts of Arizona than does Blaney-Criddle.

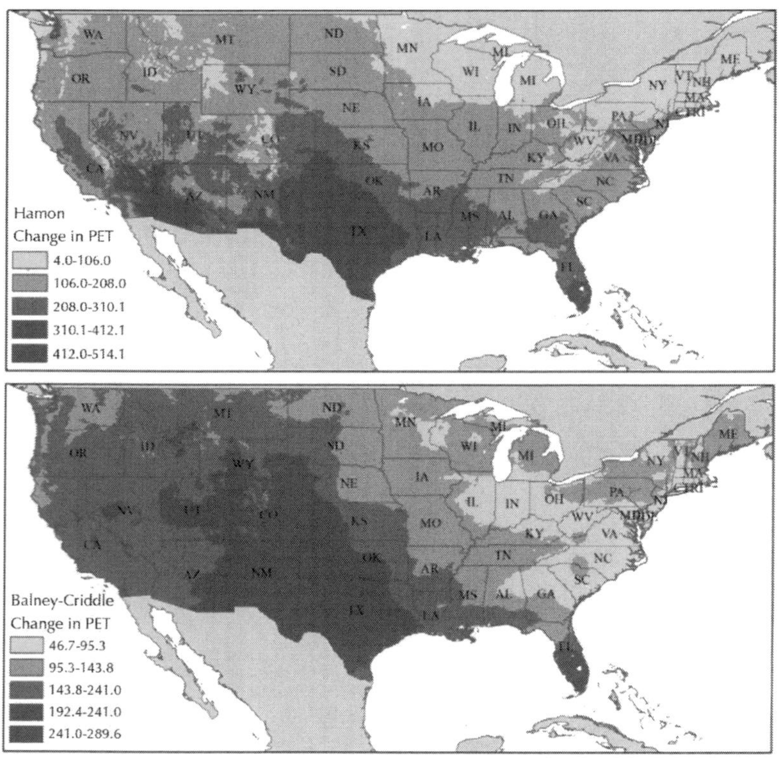

Figure 3: The effect of climate change on the Hamon estimate of PET (top) and the Blaney-Criddle estimate of PET (bottom).

Both panels are colored with 5 equal interval categories that linearly span the range of the pixel values, with areas of less increase in PET being yellow and areas of greater increase in PET being red.

Statistical Analysis

Total withdrawals were modeled as a function of total agricultural land (*A*) times the fraction of agricultural land that is irrigated (*F*) times the irrigation rate (*R*):

$$W_t = A_{tot,t} F_t R_t$$

Historical information is taken from the USDA survey. *F* and *R* were both modeled as functions of moisture deficit in 1985 and the change in moisture deficit ($\text{Deficit}_t - \text{Deficit}_{1985}$):

$$\pi_0 + \pi_1 \text{Deficit}_{1985} + \pi_2 \text{Deficit}_{\Delta}$$

where

$$\pi_0 = \mu_0 + \alpha_{t,0} + \beta_0 \text{Flood}_t$$

$$\pi_1 = \mu_1 + \alpha_{t,1} + \beta_1 \text{Flood}_t$$

The first two terms (π_0 and π_1) can be interpreted as, respectively, the slope and intercept of the relationship between moisture deficit and the response variable at a point in time. This relationship is the long-term average relationship with climate, and is the product of decades of economic decisions and irrigation policies. The third term (π_2) is the effect on the response variable of an inter-annual change in moisture deficit. It is the short term response of the agricultural sector to an interannual change in moisture deficit. All three terms are a function of a grand mean (μ), a time-specific shock (α_t), and the percent of surface irrigation (*Flood*). Surface irrigation (irrigation techniques where water is applied to the soil surface using gravity) was selected because it was believed to be causally negatively related to irrigation water-use efficiency. In this dataset proportion surface plus proportion sprinkler is approximately 1, since microirrigation is relatively uncommon. Note that the time-specific shock terms (α) will also incorporate other factors that varied over time but are not explicitly modeled here. For instance, if the steadily rising ambient CO_2 concentrations increased crop water-use efficiency over the observed 20 year time period, then one would expect the $\alpha_{t,1}$ to decline over time, all else being equal.

To account for autocorrelated errors by state, we used a repeated measured design for the error term, using Proc Mixed in SAS, version 9.2. We added terms one by one to the model, stopping when the added variables did not significantly improve the negative log-likelihood. The functional form of the regression of *F* was a logistic regression, although model fits using other functional forms that are bounded between 0 and 1 (e.g., probit) yielded similar results. For the regression of *R*, irrigation rates were log transformed to improve normality and to bound our results to the positive domain.

Future projections were calculated by assuming the fitted parameters in the regression equations, based on historical data, continue into the future. This implicitly assumes that some adaptation by producers takes place: regions that dry out, for example, adapt the agricultural practices of current comparable dry regions. Of course, hypothetically much larger water savings (or waste) are possible with more (or less) drastic change in agricultural practices, but we do not consider these more extreme possibilities in this analysis. Note that our analysis ignores the potential effect of increased CO_2 concentrations increasing crop water-use efficiency, although we highlight the importance of this issue in the Discussion section.

RESULTS

Analysis of Observed Historic Irrigation Patterns

During 1985–2005 the proportion of irrigated agricultural land was greater in areas with greater moisture deficit ($p<0.001$, Table 1; and Figure 4). In any year, states with higher moisture deficits have greater proportions of agricultural land under irrigation. This relationship represents the long-term outcome of individual farmers' decisions as a function of climate. Note that this long-term historical relationship shows adaptation to historical climate, in that dry areas are growing different crops and using different practices than wet areas. On the shorter year-to-year time scale, inter-annual changes in moisture deficit have resulted in changes to the proportion of agricultural area irrigated: higher moister deficit resulted in higher proportion of area irrigated (p = 0.03, Table 1).

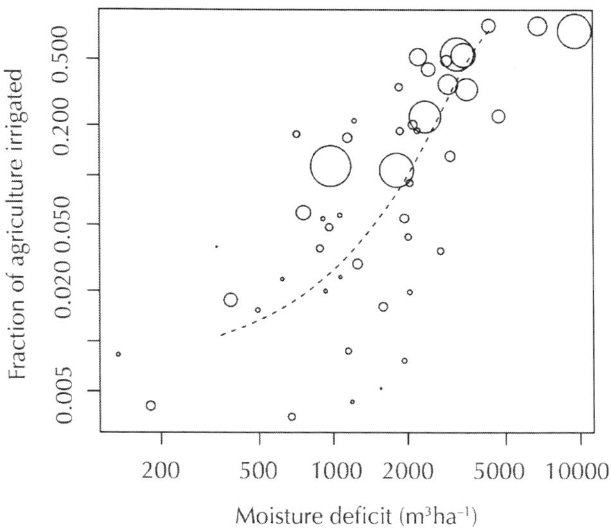

Figure 4: The fraction of agricultural land irrigated in U.S. states in 2005 as a function of moisture deficit.

Note the logarithmic scale on both axes. The size of circle is proportional to the irrigation rate (m^3ha^{-1}); states with high moisture deficit have higher irrigation rates. The fitted regression line is shown for the middle 90% of the data range.

Table 1: Regression coefficients that predict the proportion of agricultural area that is irrigated, logit transformed

Effect	Estimate	SE	P
Intercept (μ_0)	-4.4049	0.1798	<.0001
Deficit1985 (μ_1)	0.000807	0.000067	<.0001
$Deficit_\Delta$ (μ_2)	0.000371	0.000164	0.0252

Similarly, the irrigation rate (m^3 of water per hectare of irrigated land) is greater in sites with greater moisture deficit ($p<0.001$, Table 2 and Figure 4). At any point in time, states with higher moisture deficits have higher irrigation rates. Again, this relationship among different sites is the long-term outcome of a historical process of agricultural development that was influenced by the prevailing climate, and the gradient occurs despite some presumed adaptation of agricultural crops and practices to historical climate. Interestingly, the historical shift away from surface irrigation and toward sprinkler and drip irrigation has resulted in decreased irrigation rates over time ($p<0.001$, Table 2), all else being equal. In the short run on the year-to-year time scale, increases in moisture deficit are associated with increased irrigation rate, although this result is not statistically significant ($p = 0.11$, Table 2).

Table 2: Regression coefficients that predict the irrigation rate (m^3ha^{-1}), log transformed

Effect	Estimate	SE	P
Intercept (μ_0)	7.7495	0.07894	<.0001
Deficit1985 (μ_1)	0.000132	0.000037	0.0009
DeficitΔ (μ_2)	0.000118	0.000073	0.11
Fraction Flood1985 (β_1)	2.1475	0.4675	<.0001
Fraction FloodΔ (β_2)	1.3409	0.1988	<.0001

Projected Impact of Climate Change on Irrigation Area

Moisture deficit is projected to increase on average for US states under all three climate change scenarios (Figure 5A). After 2060, the increase in moisture deficit is largest in the A2 scenario, and smallest in the B1 scenario. Regardless of emissions scenario, climate change increases moisture deficit in 2090 throughout most of the contiguous US for 13 of the 16 GCMs considered [27], with the exception of the NCAR CCSM3, CSIRO MK3, and NCAR PCM models. Generally, the largest increase in moisture deficit is in the South followed by the Southwest, while the Northeast has small increases or decreases in moisture deficit. The most disagreement among climate models in the direction of moisture deficit change is in the Southwest, where some GCMs project a decrease and some an increase in moisture deficit.

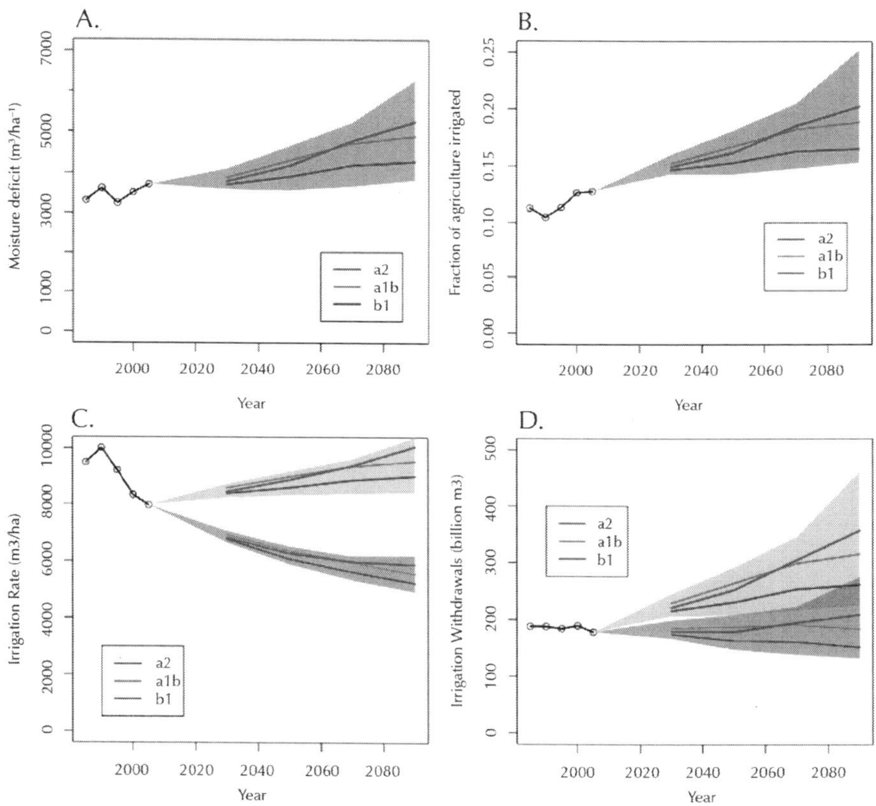

Figure 5: The effect of climate change on irrigation water use by United States agriculture.

A.) Historical and future trends in U.S. mean moisture deficit. Displayed is the irrigated area weighted average, which is most directly relevant to irrigation rate. For each climate change scenario, the projected moisture deficit at the climate of the ensemble median is shown. The grey area shows the range of fraction irrigated for climate of the 20th and 80th quintiles of the ensemble. B.) Historical and future trends in the fraction of agriculture irrigated. C.) Historical and future trends in the irrigation rate. The blue area shows the effect of climate change on irrigation rates if the current mix of irrigation equipment persists over time; the green area shows the effect of climate change if the observed (1985–2005) trend away from relatively inefficient surface irrigation continues over time. D.) Historical and future trends in total irrigation withdrawals. Note the confidence intervals of the blue (current mix of irrigation equipment) and green (decreased use of surface irrigation) areas overlap after 2030.

Given the projected increases in moisture deficit with climate change, we project that the proportion of agricultural land irrigated in the U.S. will increase (Figure 5B). Under all emissions scenarios, farmers will increasingly need to irrigate their fields as it gets drier. The total area irrigated nationally will increase by 4.5–21.9 million ha, with the uncertainty due both to differences among emissions scenarios and to differences among GCMs. The B1 emissions scenario has the smallest increase (4.5–8.7 million ha) and the A2 emissions scenario has the greatest (9.1–21.9 million ha). Note that our projection estimates new irrigated area based on observed trends, after a moderate degree of adaptation: our methodology implicitly assumes farmers in a site made drier by climate change adopt the crops and practices of current farmers in comparable climates. Depending on the state, laws governing access to water or a shortage of available water might prevent this much expansion of irrigated area from occurring [18]. Alternatively, failure of farmers to adapt to climate change as much as is assumed in our model would imply an even larger increase in irrigated area.

The states with the greatest demand for new irrigated area, in ha, are those in the South-Central U.S. (Figure 6A) and to a lesser extent the Northwest and California. There is projected to be less of an increase in the Northeast, where climate impacts on moisture deficit are projected to be smaller. The extent of expansion of irrigation is a function of the amount of greenhouse gas emissions, with the A2 scenario requiring the most expansion and the B1 scenario the least. The difference in projected new irrigation area demanded between the A2 scenario and the B1 scenario, at the ensemble median climate projections, is 6.4 million ha. While exact cumulative emissions depend on the GCM implementation of the scenarios, every 1 GtC increase in cumulative emissions increases the demand for new irrigated land in the US by roughly 7000 ha.

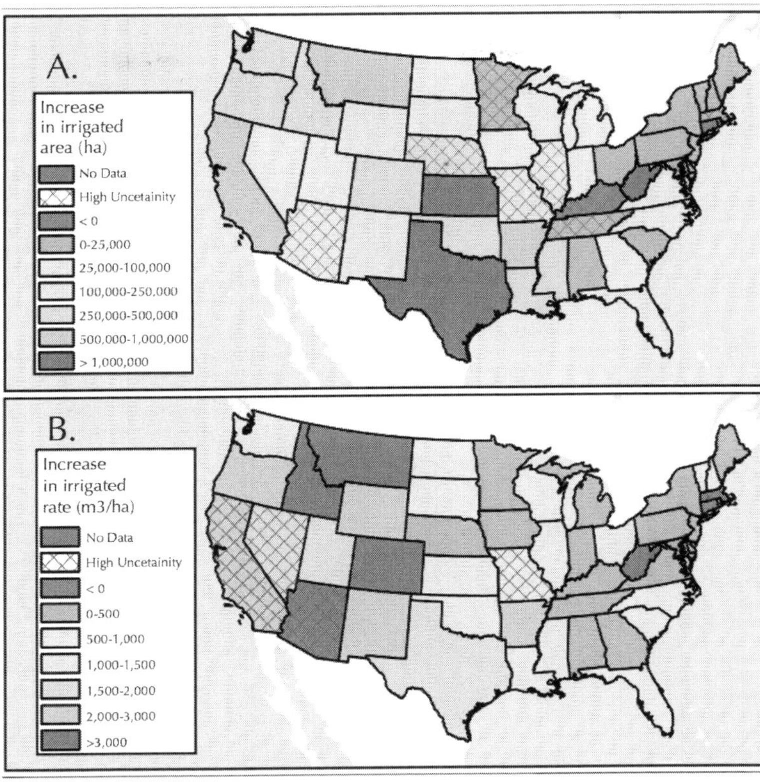

Figure 6: The effect of climate change on irrigation by states. A.) Projected increase in irrigated area by 2090 under the A1B scenario, ensemble median. Most states have an increase in irrigated area under all emission scenarios in more than 80% of the GCMs in the ensemble; those that have a decrease in some cases are marked high uncertainty. B.) Projected increase in irrigation rate by 2090 under the A1B scenario, ensemble median. Most states have an increase in irrigation rate under all emission scenarios in more than 80% of the GCMs in the ensemble; those that have a decrease in some cases are marked high uncertainty.

This expansion of agricultural area due to climate change will pose an adaptation challenge to some states that have traditionally had a very small percentage of their agricultural land irrigation. One way to see the magnitude of this challenge is to look at the fraction of agricultural water withdrawals in 2090 that would go to fields that were not irrigated in 2005 (Figure 7). In some states with currently limited irrigation systems, more than half of all irrigation withdrawals in 2090 will be for fields that were not irrigated in 2005. Changes in

Kansas (KS), Texas (TX), and Montana (MT) are particularly notable, since these states have a lot of agricultural area that is currently mostly rain-fed but will have significant expansion of their irrigation systems. Conversely, Arizona (AZ) and California (CA) already have a large fraction of their agricultural area irrigated and there is little potential for expansion to increase withdrawals in 2090.

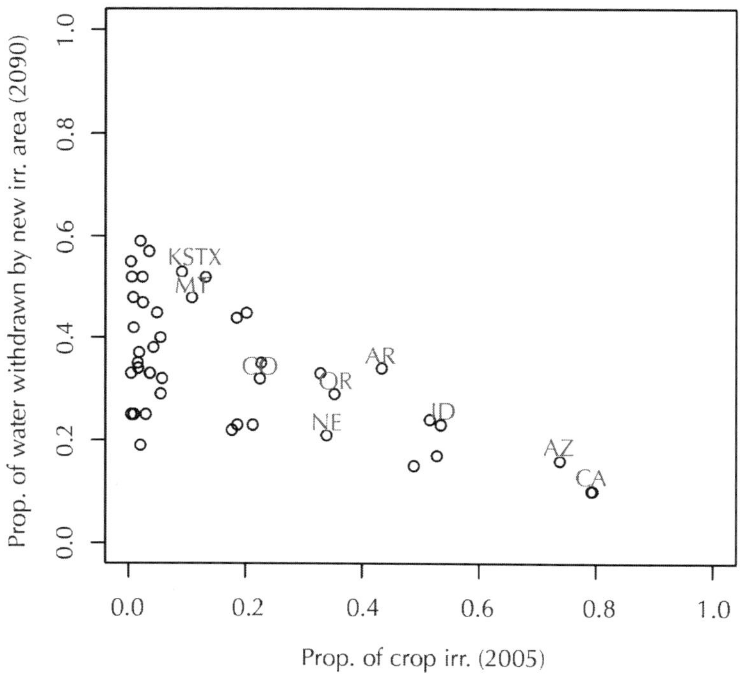

Figure 7: Currently wet states will have significant increases in irrigated area. The relationship between the proportion of agricultural land irrigation in 2005 and the predicted proportion of water withdrawals in 2090 (median A1B scenario of the GCM ensemble) that will come from fields not currently irrigated. A few states with significant agricultural area are labeled, using standard two-digit abbreviation for US states (see Figure 3). Three states are excluded from this graph, because climate change will have a net decrease on irrigated area there (Massachusetts, Connecticut, and Rhode Island).

Installing irrigation in new fields costs money, and producers will only install irrigation if they find the increased revenue from crops grown on irrigated fields worth more than the cost of irrigation. While a full analysis of the costs and benefits of installing irrigation is beyond the scope of the paper, some insights can be gained by looking at state-level averages (Figure 8). In some states that will have a large increase in proportion of agricultural area irrigated, current market value per ha of cropland is fairly high, and may be enough to support the costs of installing irrigation equipment. Conversely, producers in states with lower market value per ha may find it difficult to support the installation of irrigation equipment unless new crops or markets are found. For instance, both Arizona (AZ) and Texas (TX) are projected to have increased irrigated area in response to climate change. However, Arizona has a relatively high average market value for cropland of \$3700 ha^{-1}, driven by production of high-value crops like vegetables and horticultural products, while Texas has a relatively low average market value for cropland of \$420 ha^{-1}, primarily from production of grains and cotton.

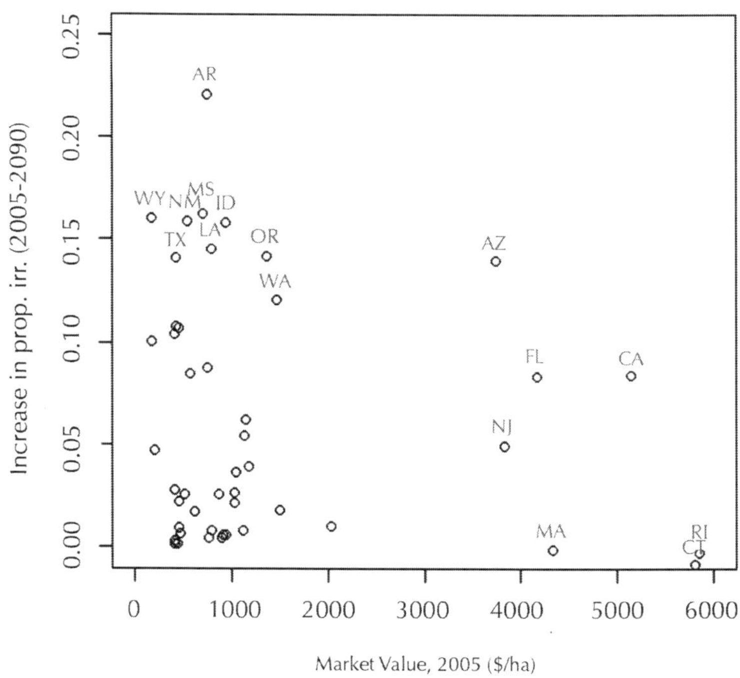

Figure 8: Market value and increase in proportion irrigated. The relationship between the average per hectare market value of cropland in 2005 and the predicted change in proportion of agricultural area irrigated (median A1B scenario of the GCM ensemble). A few states are labeled, using standard two-digit abbreviation for US states (see Figure 3).

Projected Impact of Climate Change on Irrigation Rate

Increases in moisture deficit caused by climate change will likely increase the irrigation rate (Figure 5C). Under all emissions scenarios, irrigation rates will increase with climate change from 7,963 to 8,400–10,415 m^3ha^{-1} in 2090 if the current mix of irrigation technologies stays in place. The B1 emissions scenario has the smallest rate (8,400–9,145 m^3ha^{-1}), while the A2 emissions scenario has the largest rate (9,380–10,415 m^3ha^{-1}). However, if the current trend away from surface irrigation continues at the same pace into the future, needed irrigation rates would decline from current levels to 4,911–6,205 m^3ha^{-1} in 2090 (B1 rate: 4,911–5,355 m^3ha^{-1}, A2 rate: 5,475–6,205 m^3ha^{-1}). Note that our projection only estimates irrigation rates if current observed trends continue, after a moderate degree of adaptation: our methodology implicitly assumes farmers in newly dry climates will irrigate at a rate like that of comparable dry climates. Water policy or water availability might constrain increases in the irrigation rate. Alternatively, failure of farmers to adapt to climate change as much as is assumed in our model would imply an even larger increase in the irrigation rate.

Note also that higher CO_2 levels might increase the water-use efficiency of crop plants, which would further lower the curves in Figure 5C. However, the degree to which this water-use efficiency increase from higher CO_2 concentrations will occur in real systems is unclear, depending on the water balance of the site and the plant species involved [32]–[38]. For one important crop species, wheat, increases in water-use efficiency due to higher CO_2concentrations may exceed 20% in some circumstances [37], [39], [40]. Interestingly, the $\alpha_{t,1}$terms in our model were not statistically significant and are not included in our final model, so there is no clear historical effect of rising ambient CO_2 concentrations on crop water-use efficiency.

The second challenge that climate change poses for irrigation in the US will be this increase in irrigation rates, particularly important in already arid climates. If the current mix of irrigation technologies continues into the future, the biggest increase in irrigation rate is in the Western US (Figure 6B). In places like Arizona, California, Montana, Nevada, and Idaho, farmers will have incentive to apply more water to irrigated lands. However, in some of these states (Arizona, California, and Nevada) there is high uncertainty in the predictions of the ensemble of the GCMs (Figure 6B).

Moving away from surface irrigation could generate substantial irrigation water-use efficiency gains, offsetting demand for an increase in irrigation rate in these states (Figure 9). States with the greatest potential for decrease in irrigation rate with a move away from surface irrigation are Arkansas (AR), Louisiana (LA), Arizona (AZ), Wyoming (WY), and Colorado (CO). In these states, the potential for decreased irrigation rate can more than offset the potential for irrigated area expansion due to climate change. On the other hand, some states such as Iowa and Illinois currently use very little surface irrigation, and there is little potential for decreasing irrigation rates with a move away from surface irrigation. Note also that there is considerable variability among GCMs for the Southwest, and while the median projection under all emissions scenarios is a large increase in irrigation rates in these states, a few GCMs actually project a decrease in moisture deficit and hence a decrease in irrigation rates in these states.

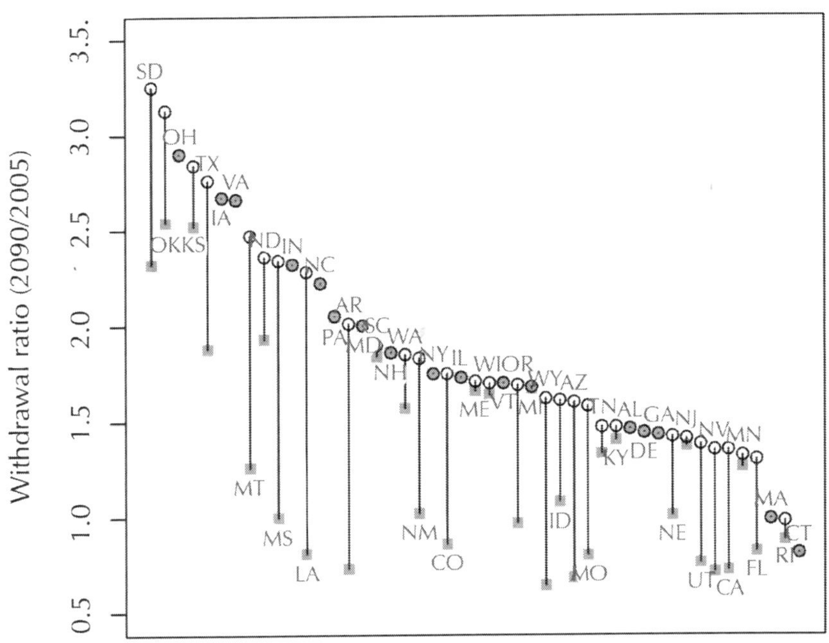

Figure 9: Withdrawal ratios by state. The ratio of irrigation withdrawals in 2090 to irrigation withdrawals in 2005 if the mix of irrigation technologies

stays the same as today (open circles) or if the current trend away from surface irrigation continues into the future (grey squares). A ratio of more than one indicates withdrawals will increase, while a value of less than one indicates withdrawals will decrease. Data points are labeled using the standard two-digit abbreviation for US states (see Figure 3), staggered so that labels do not overlap.

In the absence of gains in irrigation water-use efficiency, total water withdrawals for irrigation are likely to increase substantially (Figure 5D). This is true over all emissions pathways, but the increase in particularly striking for the A2 emissions pathway. Thus, irrigation water-use efficiency gains from switching away from surface irrigation can be an effective adaptation under moderate climate change, but cannot be wholly effective if climate change is severe. Moreover, increases in irrigation efficiency can only serve as adaptation effectively in some states (Figure 9).

DISCUSSION

Our results suggest that there are two different challenges that climate change will pose to agricultural irrigation in the US. In states in currently relatively wet climates, expansion of irrigated area is likely to be an important driver in increases in overall agricultural withdrawals. In some cases, these states have relatively little experience managing water use under scarcity, at least compared to states in the western US. For instance, many states in the eastern United States use a riparian water rights system, where all land owners along a waterbody have a right to make reasonable use of water at any time. In these states, there may currently be little regulation on the installation of a new irrigation system. Our results suggest that a key goal for these states should be developing appropriate regulations that make sure new irrigation systems have access to sufficient water without affect other water users in the area.

In contrast, in states in currently dry climate that already have a large fraction of agricultural area irrigated, our results project the predominant driver of increases in water withdrawal will be the tendency of producers to apply more irrigation water in response to droughts. Fully using technologies to increase irrigation efficiency will be a key goal for these states. The switch away from surface irrigation is included in our model, but many other forms of adaptation are possible[2], [41]. Changes in agricultural practices such as tilling and plant spacing can affect water demand, as can changes in crop varieties beyond what is implicitly assumed by our model. Many of these currently dry states are in the western US and follow a prior appropriation water rights system, where older uses of water have priority over new uses of water. New users of water may be limited by their water right in how much they can increase their irrigation volume. However, currently many states lack mechanisms for water shortages to affect the price or distribution of existing water rights to farmers, leaving them little incentive to consider low water availability in their individual decision-making [9], [42]–[45]. Farmers would have more incentive to switch to more water-efficient crops if the price of irrigation water more accurately reflects its scarcity.

Our results also highlight the importance of greenhouse gas mitigation for reducing impacts on freshwater demand. The median difference in extra demand for withdrawals in the A2 and B1 scenarios is 57 billion m^3 per year in 2090. This means that avoiding the release of 1 GtC will result in the U.S. needing roughly 65 million m^3 less irrigation water annually. Our results demonstrate that, in addition to reducing climate change's many other impacts, climate change mitigation saves irrigation water.

Our results stress that climate change will likely increase irrigation water use in the US, consistent with some previously published research [17], [18], [21], but in contrast with the results of other studies that showed mixed results depending on the GCM examined [19], [20]. At the same time, other research suggests climate change will decrease water supply, at least in some regions, either because of changes in annual supply or because of intra-annual changes in the hydrologic cycle such as changes in snowpack [19]. The exact response of agricultural producers depends on irrigation policy and law, the price of irrigation and water, and the income from crop production, among other things.

The central assumption when making climate change projections with such a statistical model is that past is prologue: how agricultural irrigation changed in response to past climate events is indicative of how it will respond in the future. While this is a reasonable assumption used by many modeling studies, readers should remember that more radical changes in water governance or water availability in the U.S. are theoretically possible, and would imply different forecasts than the ones we present.

Further research into the mechanisms that control the decisions of agricultural producers is key to evaluating the extent to which observed trends in the past two decades are indicative of the response of the agricultural system climate change. For instance, additional field-based studies of the impact of increased CO_2 on crop water use efficiency are needed to incorporate these feedbacks into mechanistic models of responses to climate change. Other factors known to affect irrigation withdrawals include the price of irrigation and income from commodity production.

Our simple statistical model provides a useful complement to more complex, mechanistic models of agricultural irrigation use. This model points out on a national scale how climate change is projected to impact agricultural water demand in different states, and how those states will likely be driven to change their agricultural practices in response to these impacts–either through increasing the area of agricultural land irrigated or through increasing the irrigation rate on already irrigated land. This information is useful to agricultural and water resource managers for informing their long-term planning, and provides insight more broadly into how changes to agricultural water demand will likely complicate water management in the future, among many other demands from other sectors.

ACKNOWLEDGMENTS

RIM is grateful for an appointment as an affiliate of the CONS Program, University of Maryland at College Park, and EHG is grateful for his affiliate appointment in the University of Washington, School of Forest Resources.

REFERENCES

1. IPCC (2007) Climate Change 2007: The Physical Science Basis; Change IPoC, editor. Cambridge, UK: Cambridge University Press.

2. Howden SM, Soussana JF, Tubiello FN, Chhetri N, Dunlop M, et al. (2007) Adapting agriculture to climate change. Proceedings of the National Academy of Sciences of the United States of America 104: 19691–19696. doi: 10.1073/pnas.0701890104
3. Schmidhuber J, Tubiello FN (2007) Global food security under climate change. Proceedings of the National Academy of Sciences of the United States of America 104: 19703–19708. doi: 10.1073/pnas.0701976104
4. Tubiello FN, Fischer GI (2007) Reducing climate change impacts on agriculture: Global and regional effects of mitigation, 2000–2080. Technological Forecasting and Social Change 74: 1030–1056. doi: 10.1016/j.techfore.2006.05.027
5. Fischer G, Shah M, Tubiello FN, van Velhuizen H (2005) Socio-economic and climate change impacts on agriculture: an integrated assessment, 1990–2080. Philosophical Transactions of the Royal Society B-Biological Sciences 360: 2067–2083. doi: 10.1098/rstb.2005.1744
6. Schlenker W, Roberts MJ (2009) Nonlinear temperature effects indicate severe damages to US crop yields under climate change. Proceedings of the National Academy of Sciences of the United States of America 106: 15594–15598. doi: 10.1073/pnas.0906865106
7. Tubiello FN, Soussana JF, Howden SM (2007) Crop and pasture response to climate change. Proceedings of the National Academy of Sciences of the United States of America 104: 19686–19690. doi: 10.1073/pnas.0701728104
8. Luedeling E, Steinmann KP, Zhang MH, Brown PH, Grant J, et al. (2011) Climate change effects on walnut pests in California. Global Change Biology 17: 228–238. doi: 10.1111/j.1365-2486.2010.02227.x
9. Rosegrant MW, Ringler C, Zhu TJ (2009) Water for Agriculture: Maintaining Food Security under Growing Scarcity. Annual Review of Environment and Resources 34: 205–222. doi: 10.1146/annurev.environ.030308.090351
10. Connor J, Schwabe K, King D, Kaczan D, Kirby M (2009) Impacts of climate change on lower Murray irrigation. Australian Journal of Agricultural and Resource Economics 53: 437–456. doi: 10.1111/j.1467-8489.2009.00460.x

11. Taylor R, Scanlon B, Doll P, Rodell M, van Beek R, et al.. (2012) Ground water and climate change. Nature Climate Change 1744: http://dx.doi.org/10.1038/nclimate1744.
12. Sun G, McNulty SG, Myers JAM, Cohen EC (2008) Impacts of Multiple Stresses on Water Demand and Supply Across the Southeastern United States. Journal of the American Water Resources Association 44: 1441–1457. doi: 10.1111/j.1752-1688.2008.00250.x
13. Palmer MA, Liermann CAR, Nilsson C, Florke M, Alcamo J, et al. (2008) Climate change and the world's river basins: anticipating management options. Frontiers in Ecology and the Environment 6: 81–89. doi: 10.1890/060148
14. Alcamo J, Florke M, Marker M (2007) Future long-term changes in global water resources driven by socio-economic and climatic changes. Hydrological Sciences Journal-Journal Des Sciences Hydrologiques 52: 247–275. doi: 10.1623/hysj.52.2.247
15. Elbakidze L (2006) Potential economic impacts of changes in water availability on agriculture in the Truckee and Carson River Basins, Nevada, USA. Journal of the American Water Resources Association 42: 841–849. doi: 10.1111/j.1752-1688.2006.tb04498.x
16. Bureau of Reclamation (2011) SECURE Water Act Section 9503(c)- Reclamation Climate Change and Water, Report to Congress. Washington, DC: US Department of the Interior, Bureau of Reclamation. 206 p.
17. Fischer G, Tubiello FN, Van Velthuizen H, Wiberg DA (2007) Climate change impacts on irrigation water requirements: Effects of mitigation, 1990–2080. Technological Forecasting and Social Change 74: 1083–1107. doi: 10.1016/j.techfore.2006.05.021
18. Schlenker W, Hanemann WM, Fisher AC (2005) Will US agriculture really benefit from global warming? Accounting for irrigation in the hedonic approach. American Economic Review 95: 395–406. doi: 10.1257/0002828053828455
19. Thomson AM, Rosenberg NJ, Izaurralde RC, Brown RA (2005) Climate change impacts for the conterminous USA: An integrated assessment - Part 5. Irrigated agriculture and national grain crop production. Climatic Change 69: 89–105. doi: 10.1007/s10584-005-3611-x

20. Rosenzweig C, Strzepek KM, Major DC, Iglesias A, Yates DN, et al. (2004) Water resources for agriculture in a changing climate: international case studies. Global Environmental Change-Human and Policy Dimensions 14: 345–360. doi: 10.1016/j.gloenvcha.2004.09.003
21. Schlenker W, Hanemann WM, Fisher AC (2007) Water availability, degree days, and the potential impact of climate change on irrigated agriculture in California. Climatic Change 81: 19–38. doi: 10.1007/s10584-005-9008-z
22. Kenny J, Barber N, Hutson S, Linsey K, Lovelace J, et al.. (2009) Estimated Use of Water in the United States in 2005. Reston, VA: U.S. Geological Survey.
23. USDA (2008) Farm and Ranch Irrigation Survey. Washington, DC: U.S. Department of Agriculture.
24. USDA (2009) US Census of Agriculture:2007. Washington, DC: United States Department of Agriculture.
25. Daly C, Neilson RP, Phillips DL (1994) A statistical-topographic model for mapping climatological precipitation over mountainous terrain. Journal of applied Meteorology 33: 140–158. doi: 10.1175/1520-0450(1994)033<0140:astmfm>2.0.co;2
26. Meehl G, Covey C, Delworth T, Latif M, McAvaney B, et al.. (2007) The WCRP CMIP3 Multimodel Dataset: A new era in climate change research. Bulletin of the American Meterological Society September: 1383–1394.
27. Girvetz EH, Zganjar C, Raber GT, Maurer EP, Kareiva P, et al.. (2009) Applied Climate-Change Analysis: The Climate Wizard Tool. Plos One 4.
28. IPCC (2000) Special Report on Emissions Scenarios. Geneva: Intergovernmental Panel on Climate Change.
29. Lu J, Sun G, McNulty SG, Amataya D (2005) A comparison of six potential evapotranspiration methods for regional use in the southeastern United States. Journal of the American Water Resources Association 41: 621–633. doi: 10.1111/j.1752-1688.2005.tb03759.x
30. Vörösmarty CJ, Federer C, Schloss AL (1998) Potential evapotranspiration functions compared on U.S. watersheds: Possible implications for global-scale water balance and terrestrial

ecosystem modelling. Journal of Hydrology 207: 147–169. doi: 10.1016/s0022-1694(98)00109-7

31. Kingston D, Todd M, Taylor R, Thompson JN, Arnell NW (2009) Uncertainty in the estimation of potential evapotranspiration under climate change. Geophysical Research Letters 36: 1–6. doi: 10.1029/2009gl040267

32. Penuelas J, Canadell JG, Ogaya R (2011) Increased water-use efficiency during the 20th century did not translate into enhanced tree growth. Global Ecology and Biogeography 20: 597–608. doi: 10.1111/j.1466-8238.2010.00608.x

33. Huang JG, Bergeron Y, Denneler B, Berninger F, Tardif J (2007) Response of forest trees to increased atmospheric CO_2. Critical Reviews in Plant Sciences 26: 265–283. doi: 10.1080/07352680701626978

34. Morgan JA, Pataki DE, Korner C, Clark H, Del Grosso SJ, et al. (2004) Water relations in grassland and desert ecosystems exposed to elevated atmospheric CO_2. Oecologia 140: 11–25. doi: 10.1007/s00442-004-1550-2

35. Wullschleger SD, Tschaplinski TJ, Norby RJ (2002) Plant water relations at elevated CO_2 - implications for water-limited environments. Plant Cell and Environment 25: 319–331. doi: 10.1046/j.1365-3040.2002.00796.x

36. Medlyn BE, Barton CVM, Broadmeadow MSJ, Ceulemans R, De Angelis P, et al. (2001) Stomatal conductance of forest species after long-term exposure to elevated CO_2 concentration: a synthesis. New Phytologist 149: 247–264. doi: 10.1046/j.1469-8137.2001.00028.x

37. Moore BD, Cheng SH, Sims D, Seemann JR (1999) The biochemical and molecular basis for photosynthetic acclimation to elevated atmospheric CO_2. Plant Cell and Environment 22: 567–582. doi: 10.1046/j.1365-3040.1999.00432.x

38. Drake BG, GonzalezMeler MA, Long SP (1997) More efficient plants: A consequence of rising atmospheric CO_2? Annual Review of Plant Physiology and Plant Molecular Biology 48: 609–639. doi: 10.1146/annurev.arplant.48.1.609

39. Manderscheid R, Weigel HJ (2007) Drought stress effects on wheat are mitigated by atmospheric CO_2 enrichment. Agronomy for Sustainable Development 27: 79–87.

40. Hunsaker DJ, Kimball BA, Pinter PJ, Wall GW, LaMorte RL, et al. (2000) CO_2enrichment and soil nitrogen effects on wheat evapotranspiration and water use efficiency. Agricultural and Forest Meteorology 104: 85–105. doi: 10.1016/s0168-1923(00)00157-x
41. Easterling WE (1996) Adapting North American agriculture to climate change in review. Agricultural and Forest Meteorology 80: 1–53. doi: 10.1016/0168-1923(95)02315-1
42. Hellegers P, Perry CJ (2006) Can irrigation water use be guided by market forces? Theory and practice. International Journal of Water Resources Development 22: 79–86. doi: 10.1080/07900620500405643
43. Tardieu H, Prefol B (2002) Full cost or "sustainability cost" pricing in irrigated agriculture. Charging for water can be effective, but is it sufficient? Irrigation and Drainage 51: 97–107. doi: 10.1002/ird.44
44. Savenije HHG, van der Zaag P (2002) Water as an economic good and demand management - Paradigms with pitfalls. Water International 27: 98–104. doi: 10.1080/02508060208686982
45. Hooker MA, Alexander WE (1998) Estimating the demand for irrigation water in the Central Valley of California. Journal of the American Water Resources Association 34: 497–505. doi: 10.1111/j.1752-1688.1998.tb00949.x

Chapter 6

Aquatic Ecosystem Response to Timber Harvesting for the Purpose of Restoring Aspen

Bobette E. Jones[1], Monika Krupa[2*], Kenneth W. Tate[2]

[1]Lassen National Forest, United States Department of Agriculture Forest Service, Susanville, California, United States of America,

[2]Department of Plant Sciences, University of California Davis, Davis, California, United States of America

ABSTRACT

The removal of conifers through commercial timber harvesting has been successful in restoring aspen, however many aspen stands are located near streams, and there are concerns about potential aquatic ecosystem impairment. We examined the effects of management-scale conifer removal from aspen stands located adjacent to streams on water quality, solar radiation, canopy cover, temperature, aquatic macroinvertebrates, and soil moisture. This 8-year study (2003–2010) involved two projects located in Lassen National Forest. The Pine-Bogard Project consisted of three treatments adjacent to Pine and Bogard Creeks: (i) Phase 1 in January 2004, (ii) Phase 2 in August 2005, and (iii) Phase 3 in January 2008. The Bailey Project consisted of one treatment adjacent to Bailey Creek in September 2006. Treatments involved whole tree removal using track-laying harvesters and rubber tire skidders. More than 80% of all samples analyzed for NO_3-N, NH_4-N, and PO_4-P at Pine, Bogard, and Bailey Creeks were below the detection limit, with the exception of naturally elevated PO_4-P in Bogard Creek. All nutrient concentrations (NO_3-N, NH_4-N, PO_4-P, K, and SO_4-S) showed little variation within streams and across years. Turbidity and TSS exhibited annual variation, but there was no significant increase in the difference between upstream and downstream turbidity and TSS levels. There was a significant decrease in stream canopy cover and increase in the potential fraction of solar radiation reaching the streams in response to the Pine-Bogard Phase 3 and Bailey treatments; however, there was no corresponding increase in stream temperatures. Macroinvertebrate metrics indicated healthy aquatic ecosystem conditions throughout the course of the study. Lastly, the removal of vegetation significantly increased soil moisture in treated stands relative to untreated stands. These results indicate that, with careful planning and implementation of site-specific best management practices, conifer removal to restore aspen stands can be conducted without degrading aquatic ecosystems.

INTRODUCTION

Trembling aspen (*Populus tremuloides* Michx.) occurs throughout North America, including the montane zone of California's Sierra Nevada and southern Cascade ranges [1]. Aspen is considered a keystone species providing crucial habitat to support a high diversity of local and landscape plant species [2], bird communities [3-5], mammals [6], and insects and other invertebrates [7]. Additionally, aspen stands provide important ecosystem services such as increased water yields and soil moisture [8], and act as natural firebreaks [9,10].

Declines in the health and distribution of aspen stands across western North America have been observed over the past century to the present day [11-14]. Much of this decline is attributable to conifer encroachment stimulated by the absence of natural fire regimes, as well as historic and current heavy browsing of aspen suckers by domestic and native herbivores [15-17]. Aspen is a clonal species and disturbance stimulates its vegetative reproduction [18]. When fire return intervals are lengthened, the disturbance of aspen and their resultant vegetative reproduction is reduced [11,19]. Additionally, fire suppression places aspen at a competitive disadvantage to shade-tolerant conifers, as aspen require sunlight and warm soil for successful regeneration [13,20,21]. The results of an aspen inventory conducted from 2000 - 2011 that assessed the current status and risk of loss of 700 live aspen stands (approximately 99 % of known stands) totaling 1,540 ha on the Eagle Lake Ranger District, Lassen National Forest, documented that 79 % of stands were at high risk of being lost. At least 45 known stands have expired with no living aspen present and no means of recruitment. Conifer encroachment was the major risk factor associated with 96 % of inventoried stands [18].

Recent studies have found that the use of commercial timber harvest techniques to remove conifers is an effective treatment for stimulating aspen regeneration [10,22,23], and that long-term success of aspen stand restoration can be achieved when this treatment is coupled with management of ungulate herbivory [16,24]. The broad-scale implementation of silvicultural treatments in this region is of concern however, because a significant number of conifer encroached aspen stands are associated with streams [18]. Timber harvest activities that occur adjacent to rivers and streams have the potential to affect soil moisture dynamics [25,26], water quality [27-33], stream temperature through increased solar radiation inputs [34-36], and aquatic communities [37-39]. However, conifer encroached aspen stands that are not released will expire and overall landscape habitat complexity and biodiversity will continue to decline. Previous studies have found that the implementation of best management practices (BMPs) during timber harvest can prevent or limit aquatic ecosystem degradation [40-42]. Such studies also indicate that, when BMPs are followed, any changes in the aquatic ecosystem that have occurred often return to pre-harvest conditions within 5 years [42-45].

The purpose of this study was to quantify the occurrence and magnitude of the impacts of management-scale prescriptive conifer removal through commercial timber harvest on aquatic habitat quality in streams adjacent to the conifer removal activities and on soil moisture dynamics within the treated aspen stands. To this end, we performed water quality, stream temperature, solar radiation, canopy cover, aquatic macroinvertebrate, and soil moisture tension monitoring, both before and after the implementation of each of four conifer removal treatments in Lassen National Forest (Figure 1). We hypothesized that conifer removal would cause: (1) an increase in nutrient concentrations, total suspended solids, and turbidity levels, (2) an increase in stream temperatures as a result of decreased canopy cover allowing more solar radiation to reach the streams, (3) a decrease in aquatic ecosystem health as indicated by macroinvertebrate metrics, and (4) an increase in soil moisture.

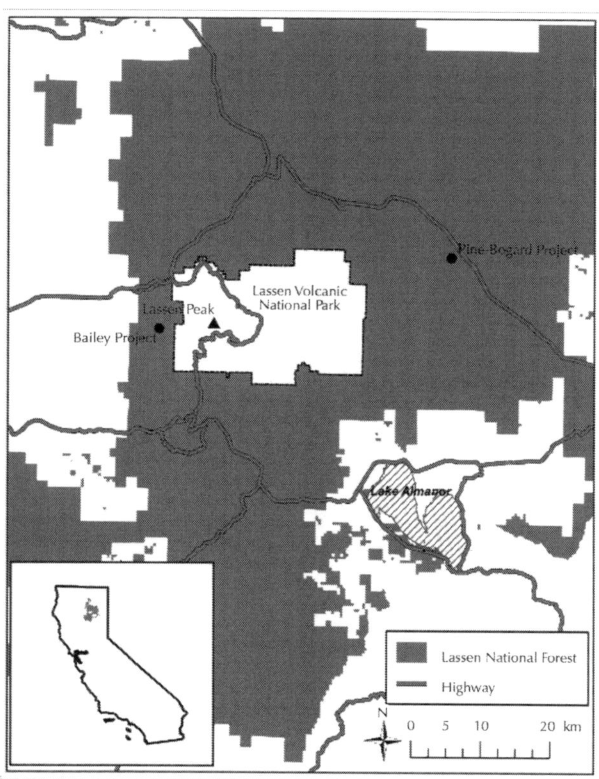

Figure 1: Pine-Bogard Project and Bailey Project locations.

MATERIALS AND METHODS

Ethics Statement

This field study was conducted in collaboration with the USDA Forest Service, and so all permissions for site access were granted and no permits were required. An Environmental Impact Statement addressing the timber harvest treatments was completed by the USDA Forest Service in accordance with National Environmental Policy Act requirements. Macroinvertebrate sampling was performed following approved protocols described by Barbour et al. [46].

Study Area

This study was carried out at two locations: (1) the confluence of Pine and Bogard Creeks (Bogard Creek is a tributary to Pine Creek) (40°34′42.66″ N, 121°05′49.18″ W) in the Eagle Lake Ranger District, which is located on the eastern side of Lassen National Forest at an elevation of approximately 1,740 m, and (2) at the South Fork of Bailey Creek (40°28′46.48″ N, 121°35′45.84″ W) in the Hat Creek Ranger District, which is located on the western side of Lassen National Forest at an elevation of approximately 1,920 m. The Pine Creek and Bailey Creek watersheds have total surface areas of approximately 591 and 154 km^2, respectively.

Vegetation in the Pine and Bogard Creek study area is dominated by Ponderosa pine (*Pinus ponderosa* Laws.), sagebrush (*Artemisia tridentata* Nutt.), and aspen. The area has slopes from 0 to 4 %, with a northeast aspect at the uppermost reaches, and level at the valley floor. Geology is dominated by basalt. Soils in the valley floor are dominated by loamy-skeletal, mixed, frigid Ultic Haploxerolls, and soils in the upper slopes are dominated by fine-loamy, mixed, frigid Pachic Ultic Argixerolls. Both Pine and Bogard Creeks are spring fed, but receive a substantial amount of snowmelt, particularly in May and June. Pine Creek has a much larger watershed area than Bogard Creek, and as a result, a larger fraction of its water is derived from snowmelt. From the months of May through September, the average width of Bogard Creek along the stream reach that was monitored in this study was approximately 1 m, and the average width of Pine Creek was approximately 3 m. During summer months, Pine Creek becomes intermittent approximately 1,500 m below its confluence with Bogard Creek. The Mediterranean climate consists of dry, warm summers, and wet, cool winters. Precipitation primarily occurs as snowfall from November through May [47]. Based on models from the PRISM Climate Group, long-term average annual precipitation is 630 mm, and ranged from 460 to 1,070 mm during the study period [48]. The primary land uses in the area are recreation (camping, hunting, snowmobiling), cattle grazing, which occurs annually from approximately June 1 through September 30, and vegetation management.

Vegetation in the Bailey Creek study area is dominated by perennial meadow herbaceous species, Lodgepole pine (*Pinus contorta* Dougl.),White fir (*Abies concolor* (Gordon & Glend) Lindley), Willow (*Salix lucida* Muhl. ssp. *lasiandra* (Benth.) E. Murray), and aspen. The area has slopes from 0 to 4 %. The geology is typified by moraine complexes with well-preserved morphology and a weakly oxidized soil zone approximately 50 cm thick. Soils are dominated by Aquolls. Bailey Creek is spring fed but receives a substantial amount of snowmelt, particularly in May and June. From the months of May through September, the average width of Bailey Creek along the stream reach that was monitored in this study was approximately 4 m. Similarly to the Pine and Bogard Creek study area, the climate is Mediterranean and precipitation primarily occurs as snowfall from November through May [47]. Based on models from the PRISM Climate Group, long-term average annual precipitation is 1,590 mm, and ranged from 810 to 1,870 mm during the study period [48]. The primary land uses in the area are recreation (camping, fishing, hunting) and vegetation management. Historically the area was grazed by livestock, but has not been grazed for more than 30 years.

Project Design

This study involved two different management-scale conifer removal projects (Figures 2 and 3). The Pine-Bogard Project consisted of conifer removal treatments in three phases: (i) Phase 1 in January 2004, (ii) Phase 2 in August 2005, and (iii) Phase 3 in January 2008. The Bailey Project consisted of conifer removal at Bailey Creek, which was implemented as one treatment in September 2006. Both project areas were chosen for restoration because the aspen stands were at high risk of loss due to extensive conifer encroachment.

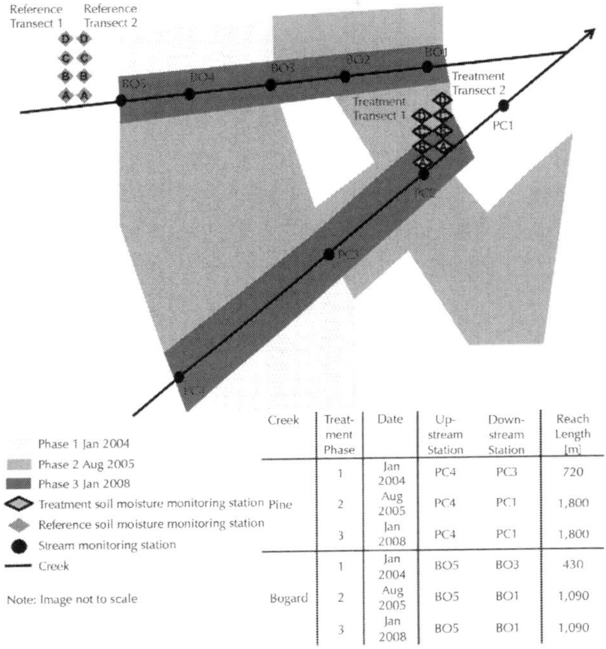

Creek	Treatment Phase	Date	Upstream Station	Downstream Station	Reach Length [m]
Pine	1	Jan 2004	PC4	PC3	720
	2	Aug 2005	PC4	PC1	1,800
	3	Jan 2008	PC4	PC1	1,800
Bogard	1	Jan 2004	BO5	BO3	430
	2	Aug 2005	BO5	BO1	1,090
	3	Jan 2008	BO5	BO1	1,090

Figure 2: Illustration of Pine-Bogard Project Phases 1-3 treatment areas and stream and soil monitoring station locations.

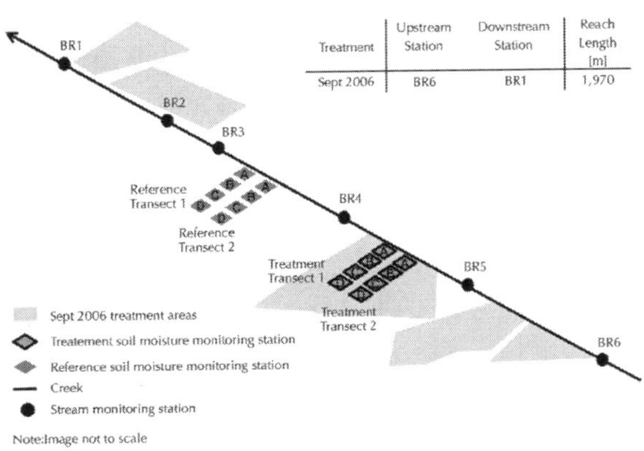

Treatment	Upstream Station	Downstream Station	Reach Length [m]
Sept 2006	BR6	BR1	1,970

Figure 3: Illustration of Bailey Project treatment areas and stream and soil monitoring station locations.

The following parameters were measured in order to understand the impacts of management-scale conifer removal on aquatic ecosystems and soil moisture dynamics: (i) soil moisture tension, (ii) stream and air temperature, (iii) stream discharge and water quality, including nitrate as N (NO_3-N), ammonium as N (NH_4-N), phosphate as P (PO_4-P), sulfate as S (SO_4-S), potassium (K), total suspended solids (TSS), and turbidity, (iv) stream canopy cover and solar radiation inputs, and (v) aquatic macroinvertebrates. Soil moisture was investigated through the establishment of discrete soil moisture tension monitoring stations along transects located within treatment and reference aspen stands (Figures 2 and 3). Treatment stands are conifer encroached aspen stands in which conifer removal took place, while reference stands are untreated conifer encroached aspen stands. Stream water quality, discharge, temperature, and aquatic macroinvertebrates, were measured at stream monitoring stations. Each stream included at least one station immediately upstream and one station immediately downstream of the treatment stands (Figures 2 and 3). Canopy cover and solar radiation measurements were made along the stream reaches located between the upstream and downstream stations for each treatment (refer to tables in Figures 2 and 3).

Conifer Removal Implementation

Pine-Bogard Phase 1 treatment occurred in January of 2004 between Pine Creek stations PC3 and PC4 and Bogard Creek stations BO3 and BO5 (Figure 2). The total treatment area for this phase was approximately 24 ha, with 720 m of Pine Creek and 430 m of Bogard Creek affected by the treatment. Timber harvest was implemented over snow with a minimum requirement of 60 cm of snow or 10 cm of frozen ground. Whole tree removal using track-laying harvesters and rubber tire skidders was used to remove conifers more than 25 m from stream. Whole tree removal is the standard practice used because it creates less slash, and over snow was used to protect the soil surface. Trees within 25 m of stream were hand felled and end-lined to reduce compaction and sedimentation within areas closest to the streams. Typically, all conifers less than 75 cm diameter at breast height (dbh) were removed, except for conifers directly contributing to stream bank stability or other site-specific benefits. Hand-felling of conifers less than 30 cm dbh occurred post-harvest.

Pine-Bogard Phase 2 treatment was designed to maximize the amount of dry season harvest in order to avoid the excessive slash that was observed during Phase 1 over snow harvest. Phase 2 treatment occurred from August 10, 2005 through September 12, 2005 between Pine Creek stations PC1 and PC4 and Bogard Creek stations BO1 and BO5 (Figure 2). The total treatment area for this phase was approximately 80 ha, with 1,800 m of Pine Creek and 1,090 m of Bogard Creek affected by the treatment. Whole tree removal using track-laying harvesters and rubber tire skidders was used to remove conifers within 4 to 40 m from stream. This distance was identified on the ground using site specific factors including slope and ground cover.

Pine-Bogard Phase 3 treatment was designed to treat the remaining areas adjacent to streams that could not be accessed during Phase 2 treatment. The treatment occurred in January 2008 between Pine Creek stations PC1 and PC4 and Bogard Creek stations BO1 and BO5 (Figure 2). The total treatment area for this phase was approximately 13 ha, with 1,800 m of Pine Creek and 1,090 m of Bogard Creek affected by the treatment. Timber harvest was implemented over snow with a minimum requirement of 60 cm of snow or 10 cm of frozen ground. Whole tree removal using track-laying harvesters and rubber tire skidders was used. A mechanical equipment boundary was delineated from the water's edge to the area with complete and continuous vegetation. No equipment was allowed to enter this area but conifers that did not contribute to streambank stability and were less than 75 cm dbh were removed through hand felling or reaching with a mechanical harvester boom.

A dry-season conifer removal project occurred in September 2006 between stations BR1 and BR6 along Bailey Creek (Figure 3). The total treatment area for this project was approximately 4.5 ha, with 560 m of the total BR1 to BR6 reach length (1,970 m) affected by the treatment. Whole tree removal using track-laying harvesters and rubber tire skidders was used between 1.5 to 90 m from stream depending upon slope and ground cover. Skid trails were designated in areas that did not contain riparian vegetation. Typically, all conifers 10 to 75 cm dbh were removed except for conifers located within 8 meters of the stream and contributing to stream bank stability. Conifers less than 50 cm dbh were hand felled. All conifers 10 cm dbh or smaller were cut, hand piled and burned outside of the aspen clone root zone to prevent damage to the root system.

Stream Discharge and Water Quality

Stream water samples were grab sampled every 2 weeks at each stream monitoring station. Pine-Bogard samples were collected from 2003 through 2010. Bailey samples were collected from 2003 through 2004, and from 2006 through 2010. Sampling was focused on both the peak flow period, which occurs in late May/early June and is the period of most rapid snowmelt, and the summer base flow period. Sampling began as early as May 19 and ended as late as October 12 at Pine and Bogard Creek stations, and began as early as June 3 and ended as late as October 7 at Bailey Creek stations. Sampling did not occur from about October to May because the streams are largely frozen and flows are low. The sampling periods for this study are similar to that of other studies performed in areas where much of the year is characterized by precipitation falling as snow and by freezing temperatures [28,32,33,44]. Historic stream discharge data from USGS Station 10359250 (1961-1978), which was located on Pine Creek near sampling location PC1 (Figure 2), indicates that the May through September sampling period generally captured at least 80 percent of annual discharge at Pine and Bogard Creeks. Readings started relatively later in the year at the Bailey Creek study area because it is located at a higher elevation than the Pine-Bogard study area and consequently snow melt occurs approximately 1 month later.

Stream discharge was measured at the same time as grab samples were collected using the area-velocity method. With this method, discharge was calculated as: mean velocity × stream width × mean stream depth [49]. A minimum of 3 stream depth readings were taken across each stream cross-section – at thalawag and at mid-point between thalawag and each streambank. Water velocity was measured at each of these three depth reading locations. A Global Waters flow meter (Global Waters Inc., Gold River, California, USA) was used to estimate velocity whenever water depth was greater than 10 cm. The float method was used whenever water depth was < 10 cm and thus could not accommodate the flow meter [49]. Velocity meter readings were collected at 0.6 of stream depth. A correction factor of 0.85 was used to adjust surface velocities when the float method was used.

Grab samples were refrigerated (4 °C) and transported to UC Davis for analysis. Nephelometric turbidity units (ntu) and pH were measured on non-filtered subsamples using standard methods SM2130 and SM4500-H^+, respectively [50]. To measure TSS, a 170 mL subsample was passed through a 0.45 μm membrane filter. The filters were dried in a desiccator until a stable weight was achieved (typically 5 to 7 days). The filter mass was measured on an analytical balance accurate to 0.001 g. The levels of TSS were then determined as the change in mass of the filter before and after filtration of the known volume of water. The concentrations of NO_3-N, NH_4-N, PO_4-P, K, and SO_4-S were measured using ion chromatography (Dionex 500x; CS12 cations; AS4A anions) on subsamples filtered through a 0.45 μm membrane filter.

Stream and Air Temperature

Stream temperature data was collected at all monitoring stations on Pine, Bogard, and Bailey Creeks (Figures 2 and 3) using Onset Optic StowAway temperature loggers. Additionally, air temperature data was collected using Onset Optic StowAway temperature loggers at one location in the Pine-Bogard Project area, and at one location at the Bailey Project area. All loggers were set to record temperature every 0.5 hours. Pine-Bogard temperature data was collected from 2003 through 2010. Bailey temperature data was collected from 2003 through 2004, and from 2006 through 2010. Loggers were deployed as early as May 19 and retrieved as late as October 12 at Pine and Bogard Creek stations, and deployed as early as June 3 and retrieved as late as October 7 at Bailey Creek stations. Daily maximum and average temperatures, as well as 7-day running average maximum and average temperatures were calculated using the recorded data.

Canopy Cover and Solar Radiation

Percent stream canopy cover was measured with a convex spherical densiometer and represents the amount of sky above a point on the stream channel which is blocked from view by vegetation. At each sample point along a given stream reach, measurements were made facing downstream, right bank, upstream, and left bank, and then the average of these values was calculated [51].

The potential fraction of solar radiation reaching each stream was estimated based on measurements collected with a solar pathfinder (Solar Pathways, Hartford, South Dakota). Solar pathfinder measurements integrate the effects of the sun's path, vegetative canopy cover, topographic shading, and stream channel aspect to estimate the potential fraction of available solar radiation (0 to 100%) reaching a site at a given latitude for each month of the year. Solar pathfinder measurements can overestimate the quantity of solar radiation reaching a site because they do not take cloud cover and atmospheric turbidity into account. A detailed description of solar pathfinders can be found in Platts et al. [52]. We concern ourselves with the months of May through September, which represent the warmest period in the region, when elevated stream water temperatures might be of concern.

There was no canopy cover or solar radiation data collected for Pine-Bogard Phase 1. Pine-Bogard Phase 2 canopy cover and solar radiation measurements were taken at the stream reach between stations BO1 and BO5, and at the stream reach between stations PC1 and PC4, before (June 2005) and after (September 2005) treatment implementation. Following Pine-Bogard Phase 3 treatment implementation, canopy cover and solar radiation measurements were taken in June 2008 at the stream reach between stations BO1 and BO5, and at the stream reach between stations PC1 and PC4. These measurements were compared to the September 2005 measurements taken following Pine-Bogard Phase 2 treatment. Measurements along Pine and Bogard Creeks were taken every 40 m during each data collection event (n= 45 and 28 for Pine and Bogard Creeks, respectively). Bailey Project canopy cover and solar radiation readings were taken at the stream reach between stations BR1 and BR6, before (September 2003) and after (July 2007) treatment implementation. Measurements along Bailey Creek were taken every 55 m (n= 36).

Aquatic Macroinvertebrates

Stream macroinvertebrate collections were made at midstream (BO3) and upstream (BO5) stations along Bogard Creek, at midstream (PC3) and upstream (PC4) stations along Pine Creek, and at midstream (BR4) and upstream (BR6) stations along Bailey Creek in June-July of 2003, 2004, 2007, 2008, and 2010. Samples were collected according to the single habitat approach for riffles described by Barbour et al. [46]. At each sample station, two sub-samples were collected along a transect perpendicular to streamflow (center-right and center-left from banks) and composited as 1 sample for analysis. Transects were established across riffle areas. All sub-samples were collected with a D-ring kick net (500 micron mesh) from a sample area of 0.09 m^2 for a sample time of 3 minutes per sample. Collections were immediately stabilized with 95 % ethanol. Taxonomic analysis was conducted at the BLM BugLab on the campus of Utah State University following the methods described by Moulton et al. [53] and Cuffney et al. [54]. Samples were taxonomically analyzed to genus and species where possible and standard metrics describing macroinvertebrate assemblage characteristics were calculated from the raw taxa data.

Soil Moisture

Soil moisture, measured as soil moisture tension in centibars, was investigated using permanently established gypsum blocks [55]. Higher values of soil moisture tension indicate drier soil conditions. Soil moisture tension measurements were taken approximately every 2 weeks from as early as May 19 to as late as October 12 from 2003 through 2010 at Pine and Bogard Creeks, and from as early as June 3 to as late as October 7 in 2004, and from 2006 through 2010, at Bailey Creek. Soil moisture tension monitoring stations were located in areas that provided a representative sample of the whole stand, excluding skid trails and landings.

At Pine and Bogard Creeks, 2 transects of 4 monitoring stations were established from near stream to uplands in the treatment area, and in a nearby reference area, in order to capture a gradient of soil moisture conditions, as illustrated in Figure 2. Gypsum blocks were permanently established at 15 and 45 cm in depth at each monitoring station (n = 16 at each depth, 8 treatment, 8 reference). These depths were chosen because they correspond to the bottom and mid-depths of herbaceous plant root mass in the Pine-Bogard and Bailey project areas. Phase 1 treatment did not affect the treatment transects. Phase 2 treatment affected stations C and D in both treatment transects, which were then compared to stations C and D in the reference transects (Figure 2). Phase 3 treatment affected stations A and B in both treatment transects, which were then compared to stations A and B in the reference transects (Figure 2).

At Bailey Creek, 2 transects of 5 monitoring stations were established from near stream to uplands in the treatment area, and in a nearby reference area, as illustrated in Figure 3. Gypsum blocks were permanently established at 15 and 45 cm in depth at each monitoring station (n = 20 at each depth, 10 treatment, 10 reference).

Data Analysis

Statistical analysis was designed to test the hypotheses that the various aquatic ecosystem parameters we measured were degraded (i.e. increased nutrient concentrations, decreased macroinvertebrate diversity) following conifer removal treatments. We used linear mixed effects (LME) and generalized linear mixed effects (GLMM) analyses to determine the occurrence and magnitude of change in each aquatic ecosystem parameter between treated/downstream and reference/upstream sample locations before versus after treatment [56-58]. A separate analysis was conducted for each parameter. The specific parameters analyzed were: NO_3-N, NH_4-N, PO_4-P, K, SO_4-S, turbidity, TSS, canopy cover, solar radiation, stream temperature, macroinvertebrate metrics (% Tolerant, % Intolerant, Shannon Diversity Index, number of families), and soil moisture tension. For soil moisture tension, separate mixed effects models were used for the 15 and 45 cm depths at (i) the Bailey transects, (ii) stations A and B of the Pine-Bogard transects, and (iii) stations C and D of the Pine-Bogard transects. The fixed effects variables in each mixed effects model were Location (downstream vs upstream; or treated vs reference area), Time (before vs after conifer removal), and the interaction of Location and Time (Time x Location). A significant interaction would indicate the presence of a treatment effect. Because there was no reference area for canopy cover and solar radiation, the mixed effects models for these parameters included only the Time variable. The two random effects variables in each mixed effects model were Sample Unit and Year, as crossed random effects. Sample Unit was used as a random term in order to account for the autocorrelation introduced by repeated measures made at each stream monitoring location or soil moisture tension sensor. Year was used as a random term to account for the annual variation in the parameters driven by environmental factors that vary across years and can be expected to affect the variables similarly across all stations (e.g. temperature and precipitation), but which were not of interest to this study. LMEs (normal distribution, homogenous variance) were used for solar radiation, canopy cover, stream temperature, and soil moisture tension analysis. GLMMs (Poisson distribution) were used for all other parameters. Conformity to the assumptions associated with these analyses was confirmed with standard diagnostic tests and graphs. All analyses were conducted with Stata/SE software [59].

RESULTS AND DISCUSSION

Water Quality

The modification of nutrient concentrations in streams through commercial timber harvest activities can alter aquatic ecosystem structure and function [34,43,45]. In particular, the levels of NO_3-N, NH_4-N, and PO_4-P are of concern because increases in these nutrients have been found to cause eutrophication [60-62]. Additionally, increases of sediment levels in streams as a result of timber harvest activities have been found to modify streambed surface conditions, decrease light penetration, and alter primary production, which can have detrimental impacts on all forms of stream biota [63].

There were no statistically significant relationships between nutrient, turbidity, and TSS levels and the Time x Location interaction ($P > 0.1$ for all analyses; $n = 275$, 362, and 315 for each analysis at Pine, Bogard, and Bailey Creeks, respectively). This is in contrast to previous studies in which nutrient concentrations and sediment levels increased substantially following timber harvest activities [30,32,33,45,64,65]. These studies however involved the clear-cutting of large portions of the watershed. Our finding are consistent with studies on partial harvesting adjacent to streams and rivers with the implementation of BMPs, in which there are limited to no effects on nutrient concentrations or sediment levels [31-33,40-42,66].

The evaluation of NO_3-N, NH_4-N, and PO_4-P across all stations and years at Pine, Bogard, and Bailey Creeks revealed extremely clean water conditions both before and after treatment implementation (Table 1). More than 80 % of all samples analyzed for NO_3-N, NH_4-N, and PO_4-P were below the detection limit, with the exception of PO_4-P in Bogard Creek, for which 14 % of samples were below the detection limit. The concentrations of nutrients measured across all stations from 2003 through 2010 are consistently below regulatory standards and similar to levels measured in unimpaired streams in the United States [67-69]. Such low levels of nutrients are characteristic of western montane forests [67,70-72].

Table 1: Results of nutrient analyses at Pine, Bogard, and Bailey Creeks

Creek	Nutrient	No. samples collected	No. samples < DL[a]	% Samples < DL	Mean of all samples	Mean of samples > DL
Pine	NO_3-N	275	232	84	0.007	0.033
	NH_4-N	275	273	99	0.026	0.137
	PO_4-P	275	249	91	0.008	0.035
Bogard	NO_3-N	362	298	82	0.007	0.028
	NH_4-N	364	364	100	0.025	--
	PO_4-P	362	49	14	0.037	0.042
Bailey	NO_3-N	315	272	86	0.005	0.021
	NH_4-N	316	293	93	0.030	0.091
	PO_4-P	315	311	99	0.006	0.041

Summary of nitrate as N (NO_3-N), ammonium as N (NH_4-N), and phosphate as P (PO_4-P) data collected for the PC1 to PC4 reach at Pine Creek (2003-2010), the BO1 to BO5 reach at Bogard Creek (2003-2010), and the BR1 to BR6 reach at Bailey Creek (2003-2004, 2006-2010).

[a] DL = detection limit; The detection limit for NO_3-N is 0.005 mg L^{-1}, for NH_4-N is 0.05 mg L^{-1}, and for PO_4-P is 0.01 mg L^{-1}.

Tables 2-7 report the TSS, turbidity, discharge, temperature, and nutrient levels observed on Pine, Bogard, and Bailey Creeks immediately upstream and downstream of treatment areas. Turbidity and TSS levels did rise and fall across years, however these changes were not associated with treatment implementation (Tables 2, 4, and 6). For example, the highest turbidity levels over the entire study period at

Bailey Creek were observed in 2004, before conifer removal took place (Table 6). High temporal variability in sediments has been detected in previous studies on the effects of timber harvesting on water quality [27,45,66]. Such variation is a result of the fact that all streams naturally gain and lose sediment along their length, and this gain-loss will vary from year to year [73,74]. A functioning stream will ultimately achieve a balance between sediment gain and loss. Lastly, although peaks in turbidity and TSS do occur, such as at station BO5 in 2004 (Table 4), it should be noted that these relatively elevated turbidity and TSS levels consistently return to normal levels in stations located immediately downstream and do not reappear in following years. This indicates that the peaks in turbidity and TSS levels, whatever their cause, are limited both spatially and temporally.

Table 2: Mean and maximum annual values of total suspended solids (TSS), turbidity, discharge and stream temperature measured at Pine Creek

					TSS		Turbidity		Discharge			Daily Water Temp	
			% Average annual		(mg L^{-1})		(ntu)		(m^3 sec^{-1})			(°C)	
Treatment	Year	n=	precipitation[a]	Station[b]	**Mean**	**Max**	**Mean**	**Max**	**Min**	**Mean**	**Max**	**Mean[c]**	**Max[d]**
Pre	2003	9	118	PC4	5.9	8.0	0.42	0.80	5	19	41	11	14
				PC3	5.1	8.6	0.37	0.58	7	12	15	11	14
				PC1	4.2	6.9	0.29	0.41	0.5	2	3	12	14
Post Phase 1	2004	10	98	PC4	3.4	7.1	0.57	1.48	3	5	10	10	13
				PC3	4.2	9.4	0.93	2.07	5	7	10	11	13
				PC1	4.1	9.4	0.87	1.77	0.2	2	2	12	15
Post Phase 1	2005	12	110	PC4	2.4	9.4	0.29	1.03	5	8	14	10	14
				PC3	4.1	11.2	0.49	2.23	7	10	15	11	14
				PC1	3.9	8.2	0.58	1.46	0.03	0.5	2	12	15
Post Phase 2	2006	9	170	PC4	2.0	4.1	0.13	0.35	10	17	44	10	12
				PC3	2.1	4.1	0.26	0.64	12	19	32	10	12
				PC1	2.9	8.8	0.51	1.58	0.07	2	3	11	14
Post Phase 2	2007	9	73	PC4	2.4	10.0	0.28	1.08	2	3	10	11	16
				PC3	2.1	4.7	0.64	3.07	2	3	5	12	16
				PC1	2.7	4.1	0.30	0.53	0.02	0.03	0.03	11	17
Post Phase 3	2008	8	76	PC4	3.3	11.8	0.81	2.42	2	3	5	13	18
				PC3	2.5	4.1	0.45	0.93	2	2	5	13	18
				PC1	2.1	4.1	0.79	1.69	0.02	0.03	0.05	10	15

Post Phase 3	2009	9	99	PC4	3.8	5.3	0.32	0.76	3	5	7	12	16
				PC3	3.9	8.2	0.38	0.75	3	5	10	13	17
				PC1	4.7	8.2	0.55	3.17	0.03	0.07	0.12	12	17
Post Phase 3	2010	7	102	PC4	3.5	5.9	0.57	1.58	5	8	12	11	15
				PC3	4.3	8.2	0.91	2.83	8	10	17	12	15
				PC1	2.8	8.2	0.80	2.94	0.07	0.2	0.7	12	16

Data was collected each year for locations immediately upstream (PC4) and downstream (PC3, PC1) of the aspen stands adjacent to Pine Creek treated during the Pine-Bogard Project.

[a] Average annual precipitation is 630 mm per year [48].

[b] PC4 = upstream of all treatment areas; PC3 = downstream of Phase 1 treatment area; PC1 = downstream of Phase 2 and 3 treatment areas.

[c] The standard error of mean daily water temperature measurements was always less than 1.3 °C.

[d] The annual average of the daily maximum stream temperatures.

Table 3: Mean annual nutrient concentrations measured at Pine Creek

				NO_3-N[b]		NH_4-N[c]		PO_4-P[d]		SO_4-S[d]		K[e]
Treatment	Year	n=	Station[a]	mg L^{-1}	%<DL[f]	mg L^{-1}	%<DL	mg L^{-1}	%<DL	**mg L^{-1}**	**%<DL**	**mg L^{-1}**
Pre	2003	9	PC4	0.0025 (0)	100	0.025 (0)	100	0.005 (0)	100	0.01 (0)	67	1.39 (0.08)
			PC3	0.003 (0.001)	88	0.025 (0)	100	0.01 (0)	75	0.01 (0)	25	1.38 (0.10)
			PC1	0.0025 (0)	100	0.025 (0)	100	0.005 (0)	100	0.01 (0)	44	1.42 (0.10)
Post Phase 1	2004	10	PC4	0.003 (0)	90	0.025 (0)	100	0.005 (0)	100	0.10 (0.01)	0	1.28 (0.05)
			PC3	0.0025 (0)	100	0.025 (0)	100	0.01 (0)	90	0.09 (0.01)	0	1.32 (0.08)
			PC1	0.0025 (0)	100	0.025 (0)	100	0.005 (0)	100	0.08 (0.02)	30	1.36 (0.07)
Post Phase 1	2005	12	PC4	0.003 (0.001)	92	0.025 (0)	100	0.01 (0)	67	0.06 (0)	0	1.51 (0.07)
			PC3	0.003 (0.001)	83	0.025 (0)	100	0.01 (0)	67	0.06 (0.01)	0	1.60 (0.14)
			PC1	0.004 (0.001)	90	0.025 (0)	100	0.01 (0)	80	0.04 (0.01)	20	1.52 (0.08)
Post Phase 2	2006	9	PC4	0.019 (0.016)	89	0.025 (0)	100	0.005 (0)	100	0.11 (0.02)	22	1.30 (0.05)
			PC3	0.0025 (0)	100	0.025 (0)	100	0.05 (0.04)	89	0.15 (0.02)	11	1.31 (0.05)

			PC1	0.0025 (0)	100	0.025 (0)	100	0.005 (0)	100	0.11 (0.02)	22	1.29 (0.06)
Post Phase 2	2007	9	PC4	0.0025 (0)	100	0.04 (0.01)	89	0.005 (0)	100	0.08 (0.02)	33	1.50 (0.10)
			PC3	0.0025 (0)	100	0.025 (0)	100	0.005 (0)	100	0.08 (0.01)	11	1.50 (0.09)
			PC1	0.0025 (0)	100	0.025 (0)	100	0.005 (0)	100	0.07 (0.02)	20	1.86 (0.41)
Post Phase 3	2008	8	PC4	0.049 (0.032)	25	0.025 (0)	100	0.01 (0)	88	0.09 (0.01)	13	1.46 (0.08)
			PC3	0.032 (0.013)	38	0.025 (0)	100	0.01 (0)	75	0.07 (0.02)	38	1.41 (0.07)
			PC1	0.016 (0.011)	50	0.025 (0)	100	0.02 (0.01)	75	0.27 (0.13)	25	1.31 (0.09)
Post Phase 3	2009	9	PC4	0.003 (0)	89	0.025 (0)	100	0.01 (0)	89	0.09 (0.01)	0	1.32 (0.05)
			PC3	0.0025 (0)	100	0.025 (0)	100	0.005 (0)	100	0.09 (0.01)	0	1.32 (0.04)
			PC1	0.003 (0.001)	86	0.025 (0)	100	0.005 (0)	100	0.07 (0.02)	29	1.33 (0.12)
Post Phase 3	2010	7	PC4	0.004 (0.001)	71	0.025 (0)	100	0.01 (0.01)	86	0.10 (0.01)	0	1.29 (0.11)
			PC3	0.008 (0.002)	43	0.04 (0.01)	86	0.01 (0)	86	0.09 (0.01)	0	1.30 (0.09)
			PC1	0.006 (0.002)	67	0.025 (0)	100	0.005 (0)	100	0.04 (0.01)	33	1.46 (0.20)

Nitrate as N (NO_3-N), ammonium as N (NH_4-N), phosphate as P (PO_4-P), sulfate as S (SO_4-S), and potassium (K) data was collected at locations immediately upstream (PC4) and downstream (PC3, PC1) of the aspen stands adjacent to Pine Creek treated during the Pine-Bogard Project.

Values in parenthesis are the standard error of the mean.

[a] PC4 = upstream of all treatment areas; PC3 = downstream of Phase 1 treatment area; PC1 = downstream of Phase 2 and 3 treatment areas.

[b] Concentrations below NO_3-N detection level were set to 0.0025 mg L^{-1} which is one-half the detection level for this analysis (0.005 mg L^{-1}).

[c] Concentrations below NH_4-N detection level were set to 0.025 mg L^{-1} which is one-half the detection level for this analysis (0.05 mg L^{-1}).

[d] Concentrations below PO_4-P and SO_4-S detection level were set to 0.005 mg L^{-1} which is one-half the detection level for this analysis (0.01 mg L^{-1}).

[e] There were no K concentrations below the detection level of 0.05 mg L^{-1}.

[f] Percent of samples below the detection limit.

Table 4: Mean and maximum annual values of total suspended solids (TSS), turbidity, discharge and stream temperature measured at Bogard Creek

					TSS		Turbidity		Discharge			Daily Water Temp	
			% Average annual		mg L^{-1}		ntu		(m^3 sec^{-1})			(°C)	
Treatment	Year	n=	precipitation[a]	Station[b]	Mean	Max	Mean	Max	Min	Mean	Max	**Mean**[c]	**Max**[d]
Pre	2003	9	118	BO5	6.9	17.1	0.76	1.57	0.8	1.4	2.0	9	15
				BO3	7.8	12.6	1.09	2.78	0.8	1.5	2.7	10	16
				BO1	5.8	11.4	1.04	2.02	0.5	1.0	1.7	10	15
Post Phase 1	2004	10	98	BO5	16.3	35.3	2.71	5.23	0.8	1.5	2.4	9	14
				BO3	7.6	20.6	1.27	3.22	0.7	1.2	1.7	10	15
				BO1	7.4	15.3	2.39	5.51	0.3	0.8	1.2	10[e]	15[e]
Post Phase 1	2005	12	110	BO5	4.5	14.7	0.66	2.44	0.8	1.2	1.7	8	14
				BO3	3.6	10.0	0.82	2.44	0.5	1.0	1.5	9	15
				BO1	3.2	6.5	0.78	3.42	0.3	0.7	1.5	10	14

Post Phase 2	2006	10	170	BO5	3.3	8.8	0.38	0.88	3.1	4.9	7.1	8	10
				BO3	3.1	5.3	0.37	0.68	3.2	4.8	6.3	8	12
				BO1	2.7	4.1	0.37	0.76	2.7	4.2	6.8	8	12
Post Phase 2	2007	9	73	BO5	10.3	24.7	2.26	4.01	0.5	0.7	0.8	10	13
				BO3	5.3	18.2	1.25	4.89	0.3	0.5	0.8	11	17
				BO1	7.6	17.1	2.06	6.41	0.2	0.3	0.5	11	15
Post Phase 3	2008	8	76	BO5	5.3	15.9	2.01	5.71	0.2	0.2	0.3	12	18
				BO3	5.1	9.4	1.47	4.82	0.2	0.2	0.3	13	19
				BO1	7.6	14.7	2.11	3.97	0.05	0.2	0.3	13	20
Post Phase 3	2009	9	99	BO5	8.0	17.7	2.55	5.57	0.2	0.2	0.3	12	17
				BO3	7.6	13.5	1.45	3.77	0.05	0.3	1.0	13	19
				BO1	8.0	22.9	1.22	2.73	0.2	0.2	0.5	13	19
Post Phase 3	2010	7	102	BO5	5.2	10.0	1.26	2.99	0.2	1.0	2.2	11	16
				BO3	6.5	11.2	2.08	3.04	0.2	0.5	1.0	12	17
				BO1	6.2	8.8	2.00	4.27	0.05	0.5	1.4	13	17

Data was collected each year for locations immediately upstream (BO5) and downstream (BO3, BO1) of the aspen stands adjacent to Bogard Creek treated during the Pine-Bogard Project.

[a] Average annual precipitation is 630 mm per year [48].

[b] BO5 = upstream of treatment areas; BO3 = downstream of Phase 1 treatment area; BO1 = downstream of Phase 2 and 3 treatment areas.

[c] The standard error of mean daily water temperature measurements was always less than 1.0 °C.

[d] The annual average of the daily maximum stream temperatures.

[e] 2004 stream temperature values at station BO1 are from station BO2 because logger at station BO1 was broken.

Table 5: Mean annual nutrient concentrations measured at Bogard Creek

				NO_3-N[b]		NH_4-N[c]		PO_4-P[d]		**SO_4-S[d]**		**K[e]**
Treatment	**Year**	**n=**	**Station[a]**	**mg L^{-1}**	**%<DL[f]**	**mg L^{-1}**	**%<DL**	**mg L^{-1}**	**%<DL**	**mg L^{-1}**	**%<DL**	**mg L^{-1}**
Pre	2003	9	BO5	0.0025 (0)	100	0.025 (0)	100	0.03 (0.01)	13	0.02 (0)	13	2.16 (0.10)
			BO3	0.0025 (0)	100	0.025 (0)	100	0.03 (0.01)	11	0.03 (0)	0	2.25 (0.07)
			BO1	0.0025 (0)	100	0.025 (0)	100	0.05 (0.02)	11	0.03 (0)	0	2.29 (0.20)
Post Phase 1	2004	10	BO5	0.003 (0.001)	80	0.025 (0)	100	0.01 (0)	40	0.18 (0.01)	0	1.99 (0.06)
			BO3	0.0025 (0)	100	0.025 (0)	100	0.01 (0)	44	0.18 (0.01)	0	1.91 (0.04)
			BO1	0.0025 (0)	100	0.025 (0)	100	0.01 (0)	44	0.16 (0.01)	0	1.95 (0.05)
Post Phase 1	2005	12	BO5	0.0025 (0)	100	0.025 (0)	100	0.04 (0)	0	0.13 (0.01)	0	2.27 (0.04)
			BO3	0.004 (0.001)	83	0.025 (0)	100	0.04 (0)	0	0.12 (0.01)	0	2.16 (0.08)
			BO1	0.003 (0)	92	0.025 (0)	100	0.04 (0)	0	0.13 (0.01)	0	2.12 (0.06)
Post Phase 2	2006	10	BO5	0.0025 (0)	100	0.025 (0)	100	0.04 (0.01)	0	0.23 (0.02)	0	2.06 (0.04)
			BO3	0.0025 (0)	100	0.025 (0)	100	0.05 (0.01)	0	0.24 (0.02)	0	2.05 (0.05)

			BO1	0.0025 (0)	100	0.025 (0)	100	0.04 (0.01)	0	0.25 (0.01)	0	2.03 (0.04)
Post Phase 2	2007	9	BO5	0.004 (0.001)	78	0.025 (0)	100	0.04 (0)	0	0.20 (0.01)	0	2.44 (0.31)
			BO3	0.005 (0.003)	89	0.025 (0)	100	0.04 (0.01)	0	0.19 (0.01)	0	2.51 (0.26)
			BO1	0.0025 (0)	100	0.025 (0)	100	0.03 (0)	0	0.20 (0.02)	0	2.34 (0.23)
Post Phase 3	2008	8	BO5	0.007 (0.003)	75	0.025 (0)	100	0.06 (0.01)	0	0.21 (0.01)	0	2.01 (0.04)
			BO3	0.013 (0.006)	63	0.025 (0)	100	0.07 (0.01)	0	0.23 (0.02)	0	2.00 (0.11)
			BO1	0.005 (0.002)	75	0.025 (0)	100	0.04 (0.01)	13	0.20 (0.01)	0	1.99 (0.11)
Post Phase 3	2009	9	BO5	0.019 (0.007)	44	0.025 (0)	100	0.03 (0.01)	22	0.17 (0.01)	0	1.86 (0.09)
			BO3	0.016 (0.006)	56	0.025 (0)	100	0.01 (0)	56	0.18 (0.01)	0	1.87 (0.12)
			BO1	0.021 (0.014)	67	0.025 (0)	100	0.01 (0)	44	0.16 (0.02)	11	1.82 (0.13)
Post Phase 3	2010	7	BO5	0.005 (0.001)	57	0.025 (0)	100	0.04 (0.01)	14	0.17 (0.02)	0	1.98 (0.10)
			BO3	0.006 (0.002)	57	0.025 (0)	100	0.05 (0.01)	14	0.19 (0.01)	0	1.90 (0.13)
			BO1	0.004 (0.001)	71	0.025 (0)	100	0.04 (0.01)	14	0.16 (0.01)	0	1.87 (0.14)

Nitrate as N (NO_3-N), ammonium as N (NH_4-N), phosphate as P (PO_4-P), sulfate as S (SO_4-S), and potassium (K) data was collected at locations immediately upstream (BO5) and downstream (BO3, BO1) of the aspen stands adjacent to Bogard Creek treated during the Pine-Bogard Project.

Values in parenthesis are the standard error of the mean.

[a] BO5 = upstream of treatment areas; BO3 = downstream of Phase 1 treatment area; BO1 = downstream of Phase 2 and 3 treatment areas.

[b] Concentrations below NO_3-N detection level were set to 0.0025 mg L^{-1} which is one-half the detection level for this analysis (0.005 mg L^{-1}).

[c] Concentrations below NH_4-N detection level were set to 0.025 mg L^{-1} which is one-half the detection level for this analysis (0.05 mg L^{-1}).

[d] Concentrations below PO_4-P and SO_4-S detection level were set to 0.005 mg L^{-1} which is one-half the detection level for this analysis (0.01 mg L^{-1}).

[e] There were no K concentrations below the detection level of 0.05 mg L^{-1}.

[f] Percent of samples below the detection limit.

Table 6: Mean and maximum annual values of total suspended solids (TSS), turbidity, discharge and stream temperature measured at Bailey Creek

					TSS		Turbidity		Discharge			Daily Water Temp	
			% Average annual		mg L^{-1}		ntu		(m^3 sec^{-1})			(°C)	
Treatment	**Year**	**n=**	**precipitation**[a]	**Station**[b]	**Mean**	**Max**	**Mean**	**Max**	**Max**	**Mean**	**Min**	**Mean**[c]	**Max**[d]
Pre	2003	6	107	BR6	4.3	6.9	0.41	0.67	59	29	8	8	10
				BR1	5.3	10.0	0.60	0.99	92	46	17	9	12
Pre	2004	8	93	BR6	2.5	7.7	1.02	3.10	58	24	5	8	11
				BR1	3.0	6.5	1.22	3.88	71	31	8	9	12
Pre	2006	7	117	BR6	2.7	6.5	0.21	0.62	54	24	7	7	10
September				BR1	3.2	5.3	0.18	0.28	65	29	7	8	11
Post	2007	8	51	BR6	1.6	3.5	0.25	0.53	19	8	3	10	13
				BR1	1.5	3.5	0.33	0.79	20	8	3	11	15
Post	2008	9	55	BR6	3.3	14.1	0.37	1.27	42	12	3	10	13
				BR1	1.7	5.9	0.52	1.08	46	15	3	11	15
Post	2009	8	69	BR6	3.7	5.3	0.56	1.95	34	17	3	9	12
				BR1	4.5	7.1	0.43	0.71	41	22	10	10	14
Post	2010	7	78	BR6	5.3	13.5	0.82	3.01	78	37	7	7	10
				BR1	6.5	8.8	0.83	1.54	134	53	19	8	11

Data was collected each year for locations immediately upstream (BR6) and downstream (BR1) of the aspen stands adjacent to Bailey Creek treated during the Bailey Project.

[a] Average annual precipitation is 1,590 mm per year [48].

[b] BR6 = upstream of treatment areas; BR1 = downstream of treatment areas.

[c] The standard error of mean daily water temperature measurements was always less than 1.9 °C.

[d] The annual average of the daily maximum stream temperatures.

Table 7: Mean annual nutrient concentrations measured at Bailey Creek

				NO_3-N[b]		NH_4-N[c]		PO_4-P[d]		**SO_4-S[e]**	**K[f]**
Treatment	**Year**	**n=**	**Station[a]**	**mg L^{-1}**	**%<DL[g]**	**mg L^{-1}**	**%<DL**	**mg L^{-1}**	**%<DL**	**mg L^{-1}**	**mg L^{-1}**
Pre	2003	6	BR6	0.004 (0.001)	67	0.06 (0.02)	50	0.005 (0)	100	3.65 (0.52)	0.91 (0.22)
			BR1	0.0025 (0)	100	0.06 (0.02)	60	0.005 (0)	100	3.19 (0.45)	0.74 (0.07)
Pre	2004	8	BR6	0.003 (0.001)	88	0.03 (0.01)	88	0.005 (0)	100	4.46 (0.59)	0.69 (0.07)
			BR1	0.004 (0.001)	88	0.025 (0)	100	0.005 (0)	100	4.22 (0.51)	0.70 (0.07)
Pre	2006	7	BR6	0.009 (0.007)	86	0.025 (0)	100	0.005 (0)	100	3.99 (0.55)	0.75 (0.06)
September			BR1	0.0025 (0)	100	0.025 (0)	100	0.005 (0)	100	3.88 (0.50)	0.77 (0.07)
Post	2007	8	BR6	0.035 (0.032)	88	0.025 (0)	100	0.005 (0)	100	5.72 (0.54)	0.79 (0.06)
			BR1	0.0025 (0)	100	0.04 (0.01)	75	0.005 (0)	100	5.35 (0.42)	0.86 (0.07)
Post	2008	9	BR6	0.0025 (0)	100	0.025 (0)	100	0.01 (0.01)	78	5.48 (0.65)	0.79 (0.05)
			BR1	0.014 (0.010)	67	0.025 (0)	100	0.005 (0)	100	5.04 (0.53)	0.84 (0.06)

Post	2009	8	BR6	0.0025 (0)	100	0.025 (0)	100	0.005 (0)	100	4.85 (0.56)	0.74 (0.06)
			BR1	0.0025 (0)	100	0.025 (0)	100	0.005 (0)	100	4.55 (0.50)	0.77 (0.07)
Post	2010	7	BR6	0.007 (0.003)	57	0.025 (0)	100	0.005 (0)	100	3.18 (0.57)	0.70 (0.07)
			BR1	0.005 (0.001)	33	0.025 (0)	100	0.005 (0)	100	2.89 (0.51)	0.75 (0.08)

Nitrate as N (NO_3-N), ammonium as N (NH_4-N), phosphate as P (PO_4-P), sulfate as S (SO_4-S), and potassium (K) data was collected at locations immediately upstream (BR1) and downstream (BR6) of the aspen stands adjacent to Bailey Creek treated during the Bailey Project.

Values in parenthesis are the standard error of the mean.

[a] BR6 = upstream of treatment areas; BR1 = downstream of treatment areas.

[b] Concentrations below NO_3-N detection level were set to 0.0025 mg L^{-1}, which is one-half the detection level for this analysis (0.005 mg L^{-1}).

[c] Concentrations below NH_4-N detection level were set to 0.025 mg L^{-1}, which is one-half the detection level for this analysis (0.05 mg L^{-1}).

[d] Concentrations below PO_4-P detection level were set to 0.005 mg L^{-1}, which is one-half the detection level for this analysis (0.01 mg L^{-1}).

[e] There were no SO_4-S concentrations below the SO_4-S detection level of 0.01 mg L^{-1}.

[f] There were no K concentrations below the K detection level of 0.05 mg L^{-1}.

[g] Percent of samples below the detection limit.

Nutrient concentrations exhibited little variation between years and between stations (Tables 3,5, and 7). With the exception of PO_4-P in Bogard Creek, the spatial and temporal consistency observed in nutrient concentrations is likely the result of the overall nutrient limitation of the creeks. The lack of variation in PO_4-P levels between stations and years at Bogard Creek (Table 5) implies that the elevated PO_4-P concentrations are a natural characteristic of Bogard Creek rather than the result of a treatment effect. As described in Section 2.1, Bogard Creek flow consists of a high fraction of sub-surface inputs, which can derive substantial quantities of P from soils and bedrock [67,75,76]. However, despite relatively elevated PO_4-P concentrations, the low levels of N in Bogard Creek likely limit the potential of P to cause eutrophication [77].

Solar Radiation and Canopy Cover

Vegetative canopy cover is a critical factor in aquatic ecosystems because it blocks solar radiation reaching stream surfaces and thus moderates water temperature [35] and influences in-stream primary production [78,79]. Additionally, vegetative canopy cover also serves as an input of nutrients and organic matter to stream systems and provides physical habitat for stream biota [34,80].

There were no statistically significant changes in canopy cover along the treatment reaches of Pine Creek (PC4 to PC1) or Bogard Creek (BO5 to BO1)in response to Pine-Bogard Phase 2 conifer removal ($P > 0.1$; $n = 74$ and 76 at Pine and Bogard Creeks, respectively). Mean canopy cover at Pine Creek was 66 % before treatment and 64 % after treatment. Mean canopy cover at Bogard Creek was 64 % before treatment and 62 % after treatment. Correspondingly, there were no significant changes in the potential fraction of solar radiation arriving at each creek ($P > 0.1$; $n = 74$ and 76 at Pine and Bogard Creeks, respectively). Mean solar radiation values for the months of May through September at Pine Creek were 31, 33, 32, 26, and 21 % before treatment and 33, 33, 33, 31, and 25 % after treatment. Mean solar radiation values for the months of May through September at Bogard Creek were 36, 38, 37, 30, and 22 % before treatment and 40, 44, 42, 31, 20 % after treatment. The lack of significant change is not surprising, given that Phase 2 conifer removal occurred primarily outside of areas immediately adjacent to Pine and Bogard Creeks (Figure 2).

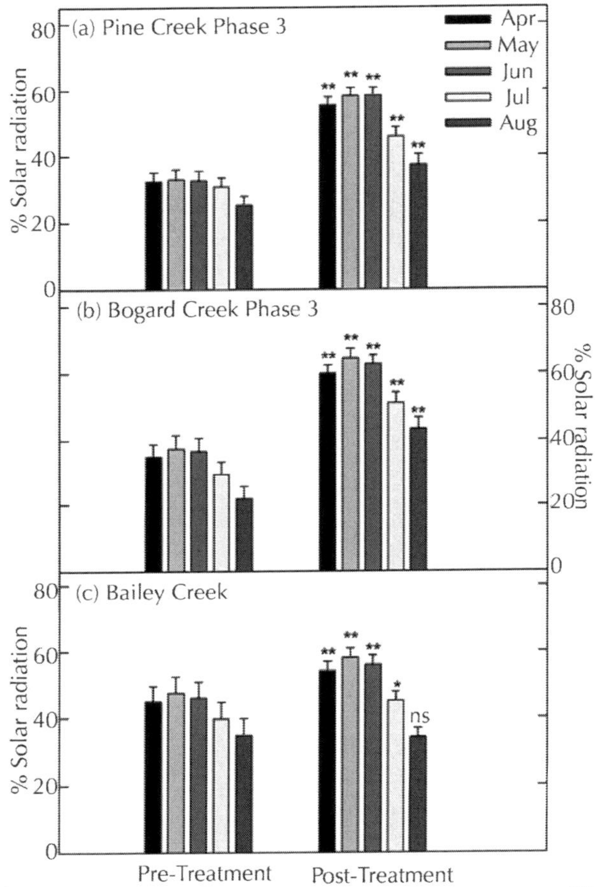

Figure 4: Mean and standard error of the potential fraction of solar radiation arriving at Pine, Bogard, and Bailey Creeks. (a) Pine Creek between stations PC4 and PC1 before (2005) and after (2008) January 2008 Phase 3 conifer removal, (b) Bogard Creek between stations BO5 and BO1 before (2005) and after (2008) January 2008 Phase 3 conifer removal, and (c) Bailey Creek between stations BR6 and BR1 before (2003) and after (2007) September 2006 conifer removal. * indicates significantly different at P<0.05. ** indicates significantly different at P<0.005.

There were significant decreases in canopy cover along the treatment reaches of Pine Creek (PC1 to PC4) and Bogard Creek (BO1 to BO5) in response to Pine-Bogard Phase 3 conifer removal ($P < 0.002$; $n = 106$ at both Pine and Bogard Creeks). Canopy cover decreased from a mean of 64 to 55 % along Pine Creek and from a mean of 64 to 39 % along Bogard Creek. Correspondingly, there was a significant increase in the potential fraction of solar radiation arriving at each reach ($P < 0.005$; $n = 75$ and 81 at Pine and Bogard Creeks, respectively).Figures 4a and b report the mean solar radiation results before and after January 2008 conifer removal. The magnitude of the increase in the potential fraction of solar radiation arriving at Pine Creek from the months of May through September ranged from 12 % in September to 25 % in June and July. The magnitude of the increase in the potential fraction of solar radiation arriving at Bogard Creek from the months of May through September ranged from 21 % in September to 27 % in June and July. The significant decreases in canopy cover and increases in solar radiation were expected, as Phase 3 conifer removal was carried out directly adjacent to Pine and Bogard Creeks.

Canopy cover significantly decreased from a mean of 64 to 55 % along the treatment reach of Bailey Creek (BR1 to BR6) in response to conifer removal ($P < 0.005$; $n = 82$). Correspondingly, there was a significant increase in the potential fraction of solar radiation arriving at Bailey Creek following treatment for the months of May through August ($P < 0.01$; $n = 82$). Figure 4creports the mean solar radiation results before and after the September 2006 conifer removal. The magnitude of the increase in the potential fraction of solar radiation ranged from 5 % in August to 11 % in June. The significant decreases in canopy cover and increases solar radiation were expected, as portions of Bailey Creek conifer removal were carried out directly adjacent to the stream.

Stream Temperature

There was no statistically significant relationship between stream temperature and the Time x Location interaction in any of the creeks ($P > 0.1$ for all analyses; $n = 276$, 334, and 305 at Pine, Bogard, and Bailey Creeks, respectively). Figures 5-7 report the 7-day running average daily maximum water temperatures observed on the upstream and downstream stations of Pine, Bogard, and Bailey Creeks. Annual stream temperature patterns appear to be driven primarily by discharge, with higher temperatures occurring in years with lower flow. The data indicates a general pattern of increased temperature from upstream to downstream stations along each creek. Although the rate of increase in stream temperature from upstream to downstream stations varied annually, Figures 5-7 confirm that it was not associated with conifer removal treatments. For example, there was no change in the rate of increase from PC4 to PC1 or from BO5 to BO1 following Phase 3 conifer removal (Figures 5 and 6). Additionally, the rate of increase at Bailey Creek was similar both before (2003-2006) and after (2007) treatment (Figure 7). It then rose in 2008, but declined again in 2009 and 2010. The lack of correspondence between the rate of increase and treatment implementation implies that fluctuations in the primary drivers of stream temperature, including groundwater inputs, hyporheic flow, air temperature and discharge [35], are the source of variation in the rates of temperature change across years.

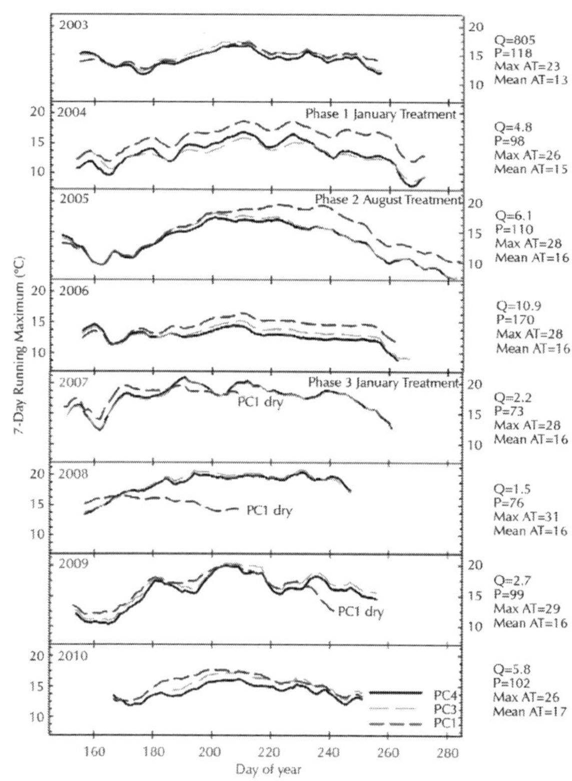

Figure 5: Seven day running average daily maximum water temperature (°C) on Pine Creek. Station PC4 was located immediately upstream of the areas

treated during Phases 1-3. Station PC3 was located immediately downstream of the areas treated during Phase 1, and station PC1 was located immediately downstream of the areas treated during Phases 2 and 3. Q = mean discharge (m^3 min^{-1}) from Jun 15 – Aug 31, P = percent of mean annual precipitation, Max AT = mean of daily maximum air temperature (°C) from Jun 15 – Aug 31, AT = mean of average daily air temperature (°C) from Jun 15 – Aug 31.

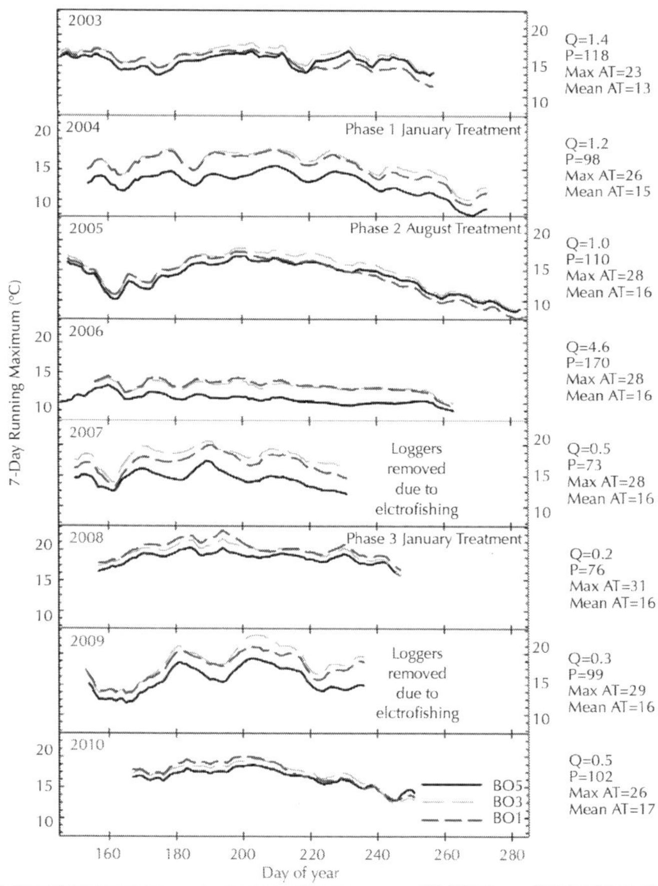

Figure 6: Seven day running average daily maximum water temperature (°C) on Bogard Creek. Station BO5 was located immediately upstream of the areas treated during Phases 1-3. Station BO3 was located immediately downstream of the areas treated during Phase 1, and station BO1 was located immediately downstream of the areas treated during Phases 2 and 3. Q = mean discharge (m^3 min^{-1}) from June 15 - August 31, P = percent of mean annual precipitation, Max AT = mean of daily maximum air temperature (°C) from

June 15 - August 31, AT = mean of average daily air temperature (°C) from June 15 - August 31. Note: the temperature logger for station BO1 in 2004 was broken, so the 2004 figure shows data collected at station BO2.

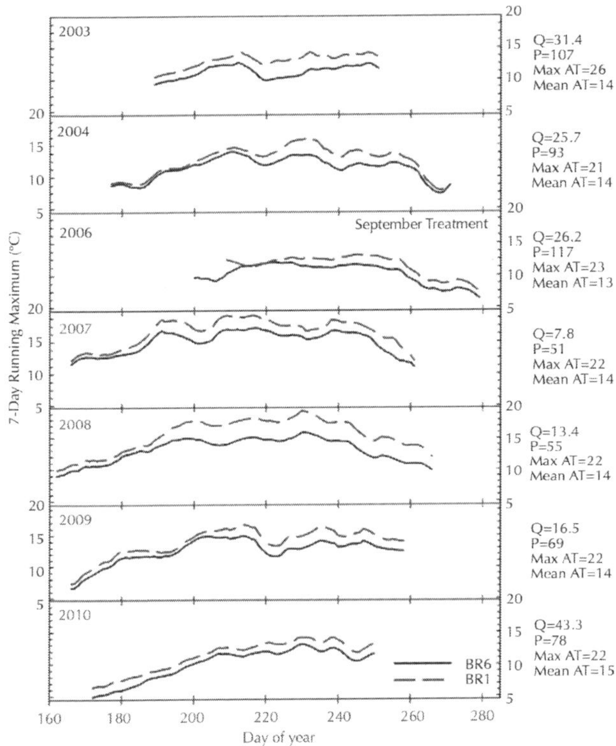

Figure 7: Seven day running average daily maximum water temperature (°C) on Bailey Creek. Stations BR6 and BR1 were located immediately upstream and downstream of the areas treated during the Bailey Project, respectively. Q = mean discharge (m^3 min^{-1}) from Jun 15 – Aug 31, P = percent of mean annual precipitation, Max AT = mean of daily maximum air temperature (°C) from Jun 15 – Aug 31, AT = mean of average daily air temperature (°C) from Jun 15 – Aug 31.

A response in stream temperature to treatment implementation was not expected to Pine-Bogard Phase 1 and Phase 2 conifer removal, as these treatments occurred primarily outside of areas immediately adjacent to the creeks (Figure 2) and Phase 2 was found to have no significant impact on canopy cover or the potential fraction solar radiation reaching the creeks. However, the Pine-Bogard Phase 3 and

Bailey treatments significantly decreased canopy cover and increased the potential fraction of solar radiation reaching the creeks, and therefore had the potential to increase downstream temperatures. There are several possible reasons for the lack of response. First, the decrease in canopy cover was minimal at Pine and Bailey Creeks (9 % and 7 % decrease, respectively). Previous studies with similar reductions in canopy cover also found no temperature response [36,66,81,82]. Second, there was still a substantial amount of canopy cover remaining at Pine Creek (55 %), Bogard Creek (39 %), and Bailey Creek (45 %) which likely provided sufficient shading to continue to moderate stream temperature. Third, at Bailey Creek, stream temperature change is likely buffered by the relatively high, cool flows that characterize the creek all season-long (Table 7). Fourth, as discussed below, the soil moisture tension data indicates that soil moisture increased as a result of vegetation removal, and this implies that there may have been an increase in water inputs to the streams that helped to buffer stream temperature from potential increases in solar radiation inputs [8,36,83]. Lastly, it is likely that the affected reach lengths at each creek were not long enough to allow for a water residence time that could result in increased temperatures. This may particularly be the case at Bogard Creek, where there was a 35 % decrease in canopy cover, and therefore an increase in temperature was expected. A shorter residence time lessens the potential for water passing through a reach to be influenced by solar radiation arriving at that reach. Chizinski et al. [81] and Hemstad et al. [82] also hypothesized that the relatively short reach lengths impacted by harvest activities adjacent to streams are one of the reasons that they found no adverse effects on aquatic ecosystem parameters.

For the purposes of aspen restoration, the scale of the timber harvest areas in this study would likely be among the largest that would be carried out should this method be applied throughout this region. This implies that the future application of these aspen restoration treatments would be unlikely to increase stream temperatures; however, projects that involve timber harvesting along longer stream reaches than those investigated in this study should monitor stream temperatures and take measures to limit potential impacts, such as leaving ample understory for shading [36,66]. In general, studies in which the clear-cutting of watersheds took place both with and without riparian buffers have shown significant increases in temperature [84,85]. In contrast, the results of our project are consistent with studies in which partial harvesting in areas adjacent to streams and rivers was carried out following BMPs, which prevented or significantly limited stream temperature changes [36,41,66,86,87].

Minimizing increases in stream temperature is a critical part of maintaining aquatic ecosystem health for stream biota. Increases in stream temperature have been found to alter macroinvertebrate abundance and community structure [43,88]. Additionally, high temperatures can substantially decrease cold-water fish populations [89]. Studies have found that the optimal temperatures for rainbow trout (*Oncorhynchus mykiss*), which is the native trout species in the region, range from 16 to 18 °C, and that the upper incipient lethal temperature for rainbow trout is approximately 25 °C [90]. Tables 2, 4, and 6 indicate mean stream temperatures did not exceed the optimal range during the course of this study, and Figures 5-7 indicate that maximum stream temperatures did not exceed the upper incipient lethal temperature. The highest temperatures observed at any of the sites throughout the course of this study occurred in Bogard Creek during the low-flow years from 2007 through 2009 (Figure 6). Despite these elevated temperatures, there was an abundance of trout in Bogard Creek during this time period, reflecting the suitability of these creeks to provide habitat for cold water fish even with timber harvesting activities [91].

Aquatic Macroinvertebrate Metrics

Table 8 reports key aquatic macroinvertebrate metrics calculated from collections made on Pine, Bogard, and Bailey Creeks. A high value of the percent of the macroinvertebrate community tolerant of pollution (% Tolerant) is an indicator of poor aquatic ecosystem health. Conversely, high values of (i) the percent of the macroinvertebrate community not tolerant of pollution (% Intolerant), (ii) of the Shannon Diversity Index, and (iii) of the number of families detected (No. Families) are indicators of good aquatic ecosystem health. There were no statistically significant relationships between these metrics and the Time x Location interaction ($P > 0.1$; $n = 10$ for the analysis of each parameter at each creek).

Table 8: Aquatic macroinvertebrate metrics for Pine, Bogard, and Bailey Creek samples collected in June-July of 2003-2004, 2007-2008, and 2010

		Midstream station[b]					Upstream station[c]				
Creek	**Metric[a]**	**2003**	**2004**	**2007**	**2008**	**2010**	**2003**	**2004**	**2007**	**2008**	**2010**
Pine	No. Families	17	17	19	23	26	16	11	21	25	21
	Shannon D.I.	2.34	2.62	1.13	1.40	1.84	2.37	1.68	1.02	1.50	1.81
	% Tolerant	0.2	0	0	0	0.1	0	0	0	0	0
	% Intolerant	16	25	7	8	10	16	14	4	7	8
Bogard	No. Families	17	17	15	31	22	12	20	21	20	23
	Shannon D.I.	2.83	2.09	2.76	2.69	2.67	2.00	2.22	2.23	1.91	2.25
	% Tolerant	0	0	0	0	0	0	0	0	0	0
	% Intolerant	24	13	21	18	22	29	11	17	7	11
Bailey	No. Families	11	12	12	15	21	13	10	12	18	17
	Shannon D.I.	2.03	2.19	1.96	2.49	2.78	2.31	2.01	2.08	2.59	2.92
	% Tolerant	0	0	0	0	0	0	0	0	0	0
	% Intolerant	50	37	27	30	53	23	52	33	42	38

[a] No. Families = number of families; Shannon D.I. = Shannon Diversity Index; % Tolerant = percent of the macroinvertebrate community tolerant of pollution; % Intolerant = percent of the macroinvertebrate community intolerant of pollution.

[b] Midstream stations were PC3, BO3, and BR4 at Pine, Bogard, and Bailey Creeks, respectively.

[c] Upstream stations were PC4, BO5, and BR6 at Pine, Bogard, and Bailey Creeks, respectively.

The results indicate healthy in-stream habitat conditions across all three creeks throughout the course of this study. The value of % Tolerant was zero in 93 % of the samples collected across all stations and years. The highest value of % Tolerant was 0.2 % measured in 2003 at midstream station PC3 prior to treatment implementation. At each creek, the Shannon Diversity Index, No. Families, and % Intolerant varied from year to year, but variation of similar magnitudes occurred at all stations (upstream and midstream), and the annual changes were not indicative of a decline in aquatic ecosystem health in response to conifer removal treatments. These results parallel the findings of the Kreutzweiser et al. [41] study, which found that most changes in macroinvertebrate community metrics before versus after partial timber harvesting adjacent to two watershed streams were similar in magnitude to the changes detected at the reference watershed and concluded that the timber harvest activities studied did not cause degradation of macroinvertebrate community structure and function. Similarly, studies by Chizinski et al. [81] and Gravelle et al. [39] found high annual variation in macroinvertebrate metrics but no treatment effect in response to partial timber harvesting or clear cutting activities. Strong temporal variability in macroinvertebrate metrics has been commonly observed in aquatic ecosystems, particularly in response to high annual variation in rainfall, which is characteristic of Mediterranean climates [92-94]. This implies that the variability detected in this study reflects site specific responses to annual climatic variation rather than the effects of conifer removal treatments.

Although strong temporal variability can mask treatment effects [95,96], the apparent lack of response of macroinvertebrate metrics to conifer removal treatments is likely the result of a lack of treatment effect on key stream characteristics. Timber harvest activities affect stream macroinvertebrate communities through several mechanisms. Aquatic macroinvertebrate community structure and function has been found to be degraded by increases in stream temperature due to reductions in canopy cover, by nutrient enrichment due to soil disturbance, and by fine inorganic sediment loading following timber harvesting [34,97,98]. Previous studies in which temperatures, sediments, and algae increased following timber harvest activities found negative impacts on the stream macroinvertebrate communities [38,78,99]. In contrast, stream parameters showed little to no response to the timber harvest activities carried out in the studies by Chizinski et al. [81],

Hemstad et al. [82], and Kreutzweiser et al. [41]. Correspondingly, there were no adverse effects detected on stream biota. In this study, stream temperature, nutrient concentrations, and TSS and turbidity levels did not respond to treatment implementation. As a result, the lack of response of macroinvertebrate metrics to conifer removal treatments is not surprising.

Soil Moisture Tension

The statistical analysis results show that the difference between soil moisture tension in treatment and reference transects increased significantly at both depths in response to the Bailey Project ($P < 0.001$; $n = 725$ and 690 at the 15 and 45 cm depths, respectively) and in response to Phase 2 ($P < 0.001$; $n = 666$ and 662 at the 15 and 45 cm depths, respectively) and Phase 3 ($P < 0.001$; $n = 677$ and 678 at the 15 and 45 cm depths, respectively) of the Pine-Bogard Project. In response to Phase 2 conifer removal, the difference between treatment and reference soil moisture tension increased by 48 and 38 kPa at the 15 and 45 cm depths, respectively. Similarly, in response to Phase 3 conifer removal, the difference between treatment and reference soil moisture tension increased by 31 and 87 kPa at the 15 and 45 cm depths, respectively. Lastly, in response to Bailey Project conifer removal, the difference between treatment and reference soil moisture tension increased by 47 and 67 kPa at the 15 and 45 cm depths, respectively.

The soil moisture tension data illustrated in Figures 8-10 corroborates the results of the statistical analysis and shows that soil moisture in treatment transects increased relative to soil moisture in reference transects at both soil depths. At the Pine-Bogard location, soil moisture tension at both depths was lower (i.e. soil moisture was higher) in the treatment transects than in the reference transects throughout the course of this study (Figures 8 and 9). However, the reference transects exhibited a higher rate of drying relative to treatment transects following both Phase 2 and Phase 3 conifer removal. This trend is particularly evident in the lowest precipitation years from 2007 through 2009 following Phase 2 treatment (Figure 8). During this period, soil moisture tension in reference transects increased (i.e. soil moisture decreased) at a rapid rate relative to previous years, while soil moisture tension in treatment transects showed relatively little response to the drought conditions. At

Bailey Creek, soil moisture tension at both depths was higher (i.e. soil moisture was lower) in the treatment transects than in the reference transects prior to treatment implementation (Figure 10). Following conifer removal in September 2006 however, soil moisture tension was lower (i.e. soil moisture was higher) in the treatment transects than in the reference transects.

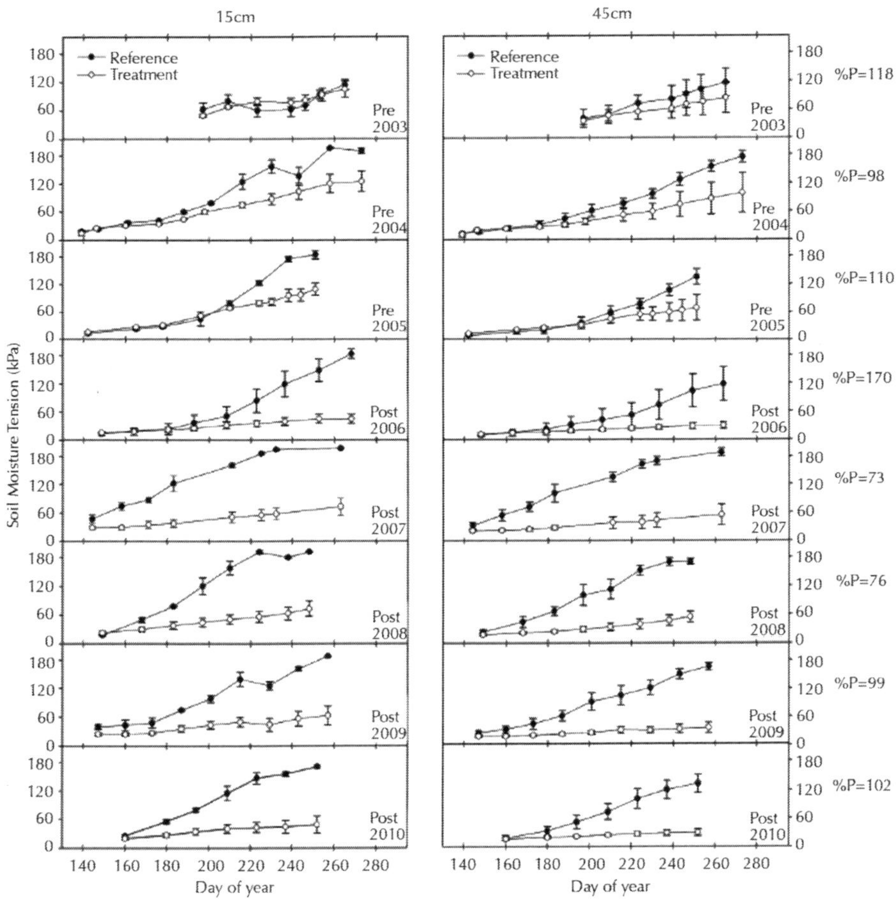

Figure 8: Mean and standard error of soil moisture within the Pine-Bogard Phase 2 August 2005 treatment area. Measurements were made at the 15 and 45 cm depths at Stations C and D of the reference and treatment transects. % P = percent of mean annual precipitation. Lower values of soil moisture tension correspond to wetter soils.

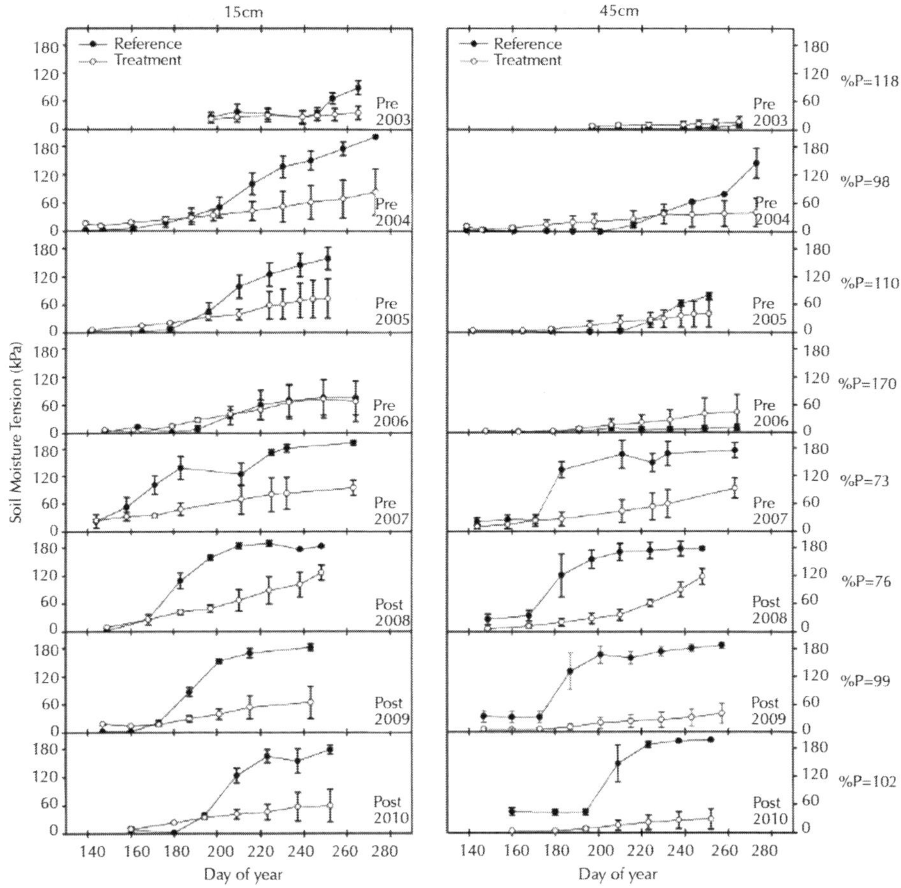

Figure 9: Mean and standard error of soil moisture within the Pine-Bogard Phase 3 January 2008 treatment area. Measurements were made at the 15 and 45 cm depths at Stations A and B of the reference and treatment transects. % P = percent of mean annual precipitation. Lower values of soil moisture tension correspond to wetter soils.

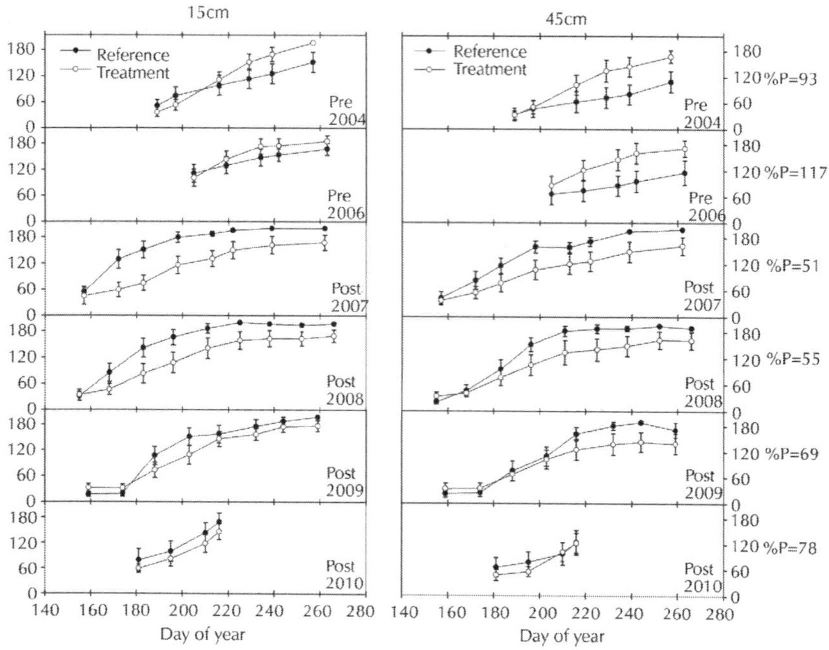

Figure 10: Mean and standard error of soil moisture at the Bailey Creek reference and treatment transects. Measurements were made at the 15 and 45 cm depths. % P = percent of mean annual precipitation. Lower values of soil moisture tension correspond to wetter soils.

The greater retention of soil moisture within treatment transects is most likely the result of the removal of vegetation causing a reduction in transpiration [83], which sustains high soil moisture levels into the dry season. Additionally, the increase in soil moisture could be the result of a decrease in snow sublimation and increase in snow water equivalent (i.e. snow accumulation) that occurs when forest canopy cover decreases [100-103]. Increases in soil moisture in response to timber harvesting have been observed in multiple studies, though, in the long-term, soil moisture levels decrease as vegetation regrows [25,26]. Within conifer-encroached aspen stands however, high soil moisture levels have the potential to be sustained if the site is successfully recolonized by aspen, because mature conifer forests use more water than mature aspen forests [8,104] and because, as a result of higher canopy interception by confers, conifer forests have lower snow water equivalent than deciduous forests [8,103,105,106]. Elevated soil moisture relative to conifer stands is one of the potential factors causing high diversity and productivity in the herbaceous understory of aspen stands [2,107].

CONCLUSIONS

The results of this study suggest that, with careful consideration of site specific conditions and implementation of appropriate best management practices, conifer removal through commercial timber harvesting for the purposes of restoring aspen stands in the southern Cascades can be implemented without degrading the aquatic ecosystem parameters measured. More than 80 % of all stream water samples analyzed for NO_3-N, NH_4-N, and PO_4-P at Pine, Bogard, and Bailey Creeks were below the detection limit, with the exception of PO_4-P in Bogard Creek, in which concentrations were elevated due to the spring source of the streamwater. There was no significant increase in the difference between upstream and downstream turbidity, TSS, NO_3-N, NH_4-N, PO_4-P, K, and SO_4-S levels before versus after treatment. There was a significant decrease in canopy cover and increase in the potential fraction of solar radiation reaching the creeks in response to the Pine-Bogard Phase 3 and Bailey treatments; however, there was no corresponding increase in the difference between upstream and downstream temperatures. Macroinvertebrate metrics confirmed the water quality results, with the highest level of % Tolerant species being 0.2 % at Pine Creek prior to treatment implementation. Lastly, soil moisture tension measurements indicate that there was a significant increase in soil moisture in treated aspen stands relative to untreated stands. Although a large body of literature exists in which timber harvest activities were found to impair water quality and aquatic ecosystem functions, the findings of this study concur with recent studies in which the partial harvesting of areas near streams and rivers with the implementation of BMPs was conducted without the degradation of aquatic ecosystems.

ACKNOWLEDGMENTS

We thank the many field assistants who collected all the valuable data, especially, Shannon Cler and Sarah Hubert; Natalie Stoddard, Betsy

Huang, and Yukako Sado for sample analysis and data management; Melanie McFarland, Tom Rickman, Coye Burnett, Dave Burton, and Alfred Vazquez for their contributions, support, advice, and encouragement throughout this project; and the Quincy Library Group for their support.

REFERENCES

1. Perala DA (1991) *Populus tremuloides* Michx. In: RM BurnsBH Honkala. Handbook 654, Silvics of North America, Volume 2, Hardwoods. Washington DC: U.S. Department of Agriculture. pp. 555-569.
2. Kuhn TJ, Safford HD, Jones BE, Tate KW (2011) Aspen (Populus tremuloides) stands and their contribution to plant diversity in a semiarid coniferous landscape. Plant Ecology 212: 1451-1463. doi:10.1007/s11258-011-9920-4.
3. Johns BW (1993) The influence of grove size on bird species richness in aspen parklands. Wilson Bulletin 105: 256-264.
4. Griffis-Kyle KL, Beier P (2003) Small isolated aspen stands enrich bird communities in southwestern ponderosa pine forests. Biological Conservation 110: 375-385. doi:10.1016/S0006-3207(02)00237-9.
5. Richardson TW, Heath SK (2004) Effects of conifers on aspen-breeding bird communities in the Sierra Nevada. Trans Western Sect Wildlife Soc 40: 68-81.
6. Debyle NV (1985) Wildlife. In: NV DeByleRP Winokur. Aspen: ecology and management in the western United States. General Technical Report RM-119. Fort Collins, CO: U.S. Department of Agriculture, Forest Service, Rocky. Mountain Forest and Range Experiment Station. pp. 135-152.
7. Jones JR, DeByle NV, Bowers DM (1985) Insects and other invertebrates. In: NV DeByleRP Winokur. Aspen: ecology and management in the western United States. General Technical Report RM-119. Fort Collins, CO: U.S. Department of Agriculture, Forest Service, Rocky . Mountain Forest and Range Experiment Station . pp. 107-114.

8. LaMalfa EM, Ryle R (2008) Differential snowpack accumulation and water dynamics in aspen and conifer communities: implications for water yield and ecosystem function. Ecosystems 11: 569-581. doi:10.1007/s10021-008-9143-2.

9. Fechner GH, Barrows HS (1976) Aspen stands as wildfire fuel breaks. Eisenhower Consortium Bulletin 4. Available: http://digitalcommons.usu.edu/aspen_bib/5029/ . Accessed 5 December 2013.

10. Kilpatrick S, Clause D, Scott D (2003) Aspen response to prescribed fire, mechanical treatments, and ungulate herbivory. In: PN OmiLA Joyce. Fire, fuel treatments, and ecological restoration: conference proceedings 16-18 April 2002. Fort Collins, CO: U.S. Department of Agriculture, Forest Service, Rocky. Mountain Forest and Range Experiment Station. pp. 93-102.

11. Smith AE, Smith FW (2005) Twenty-year change in aspen dominance in pure aspen and mixed aspen/conifer stands on the Uncompahgre Plateau, Colorado, USA. Forest Ecology and Management 213: 338-348. doi:10.1016/j.foreco.2005.03.018.

12. Rogers P (2002) Using Forest Health Monitoring to assess aspen forest cover change in the southern Rockies ecoregion. Forest Ecology and Management 155: 223-236. doi:10.1016/S0378-1127(01)00560-6.

13. Calder WJ, Horn KJ, St Clair SB (2011) Conifer expansion reduces the competitive ability and herbivore defense of aspen by modifying light environment and soil chemistry. Tree Physiol 31: 582-591. doi:10.1093/treephys/tpr041. PubMed: 21602559.

14. Rehfeldt GE, Ferguson DE, Crookston NL (2009) Aspen, climate, and sudden decline in western USA. Forest Ecology and Management 258: 2353-2364. doi:10.1016/j.foreco.2009.06.005.

15. Bartos DL, Campbell RB Jr. (1998) Decline of quaking aspen in the Interior West: examples from Utah. Rangelands 20: 17-24.

16. Jones BE, Lile DF, Tate KW (2009) Effect of simulated browsing on aspen regeneration: implications for restoration. Rangeland Ecology and Management 62: 557-563 10.2111/.1/REM-D-09-00082.1. doi: 10.2111/.1/rem-d-09-00082.1

17. Pierce AD, Taylor AH (2010) Competition and regeneration in quaking aspen-white fir (Populus tremuloides-Abies concolor)

forests in the Northern Sierra Nevada, USA. Journal of Vegetation Science 21: 507-519. doi:10.1111/j.1654-1103.2009.01158.x.

18. Shepperd WD, Rogers PC, Burton D, Bartos DL (2006) Ecology, biodiversity, management, and restoration of aspen in the Sierra Nevada, General Technical Report RMRS-GTR-178. Fort Collins, CO: U.S. Department of Agriculture Forest Service, Rocky Mountain Forest and Range Experiment Station. p. 122.
19. Smith EA, O'Loughlin D, Buck JR, St Clair SB (2011) The influences of conifer succession, physiographic conditions and herbivory on quaking aspen regeneration after fire. Forest Ecology and Management 262: 325-330. doi:10.1016/j.foreco.2011.03.038.
20. Doucet R (1989) Regeneration silviculture of aspen. Forestry Chronicle 65: 23-27. doi: 10.5558/tfc65023-1
21. Fraser EC, Lieffers VJ, Landhausser SM, Frey BR (2002) Soil nutrition and temperature as drivers of root suckering in trembling aspen. Canadian Journal Forest Research-Revue Canadienne de Recherche Forestiere 32: 1685-1691. doi:10.1139/x02-080.
22. Jones BE, Rickman TH, Vazquez A, Sado Y, Tate KW (2005) Removal of encroaching conifers to regenerate degraded aspen stands in the Sierra Nevada. Restoration Ecology 13: 373-379. doi:10.1111/j.1526-100X.2005.00046.x.
23. Shepperd WD (2001) Techniques to restore aspen forests in the western U.S. Trans Western Sect Wildlife Soc 40: 52-60.
24. White CA, Olmsted CE, Kay CE (1998) Aspen, elk, and fire in the Rocky Mountain national parks of North America. Wildlife Society Bulletin 26: 449-462.
25. Adams PW, Flint AL, Fredriksen RL (1991) Long-term patterns in soil-moisture and revegetation after a clear-cut of a douglas-fir forest in Oregon. Forest Ecology and Management 41: 249-263. doi:10.1016/0378-1127(91)90107-7.
26. Gray AN, Spies TA, Easter MJ (2002) Microclimatic and soil moisture responses to gap formation in coastal Douglas-fir forests. Canadian Journal Forest Research-Revue Canadienne de Recherche Forestiere 32: 332-343. doi:10.1139/x01-200.
27. Croke JC, Hairsine PB (2006) Sediment delivery in managed forests: a review. Environmental Reviews 14: 59-87. doi:10.1139/a05-016.

28. Tremblay Y, Rousseau AN, Plamondon AP, Levesque D, Prevost M (2009) Changes in stream water quality due to logging of the boreal forest in the Montmorency Forest, Quebec. Hydrological Processes 23: 764-776. doi:10.1002/hyp.7175.
29. Bolda KS, Meyers WJ (1997) Conducting a long-term water quality monitoring project: A case study on the McCloud River. California. Journal- Journal of Soils and Water Conservation 52: 49-54.
30. Klein RD, Lewis J, Buffleben MS (2012) Logging and turbidity in the coastal watersheds of northern California. Geomorphology 139: 136-144. doi: 10.1016/j.geomorph.2011.10.011
31. Litschert SE, MacDonald LH (2009) Frequency and characteristics of sediment delivery pathways from forest harvest units to streams. Forest Ecology and Management 259: 143-150. doi:10.1016/j.foreco.2009.09.038.
32. Karwan DL, Gravelle JA, Hubbart JA (2007) Effects of timber harvest on suspended sediment loads in Mica Creek, Idaho. Forest Science 53: 181-188.
33. Gravelle JA, Ice G, Link TE, Cook DL (2009) Nutrient concentration dynamics in an inland Pacific Northwest watershed before and after timber harvest. Forest Ecology and Management 257: 1663-1675. doi:10.1016/j.foreco.2009.01.017.
34. Campbell IC, Doeg TJ (1989) Impact of timber harvesting and production on streams - a review. Australian Journal of Marine and Freshwater Research 40: 519-539. doi:10.1071/MF9890519.
35. Poole GC, Berman CH (2001) An ecological perspective on in-stream temperature: Natural heat dynamics and mechanisms of human-caused thermal degradation. Environ Manage 27: 787-802. doi:10.1007/s002670010188. PubMed: 11393314.
36. Gravelle JA, Link TE (2007) Influence of timber harvesting on headwater peak stream temperatures in a northern Idaho watershed. Forest Science 53: 189-205.
37. Herlihy AT, Gerth WJ, Li J, Banks JL (2005) Macroinvertebrate community response to natural and forest harvest gradients in western Oregon headwater streams. Freshwater Biology 50: 905-919. doi:10.1111/j.1365-2427.2005.01363.x.

38. Reid DJ, Quinn JM, Wright-Stow AE (2010) Responses of stream macroinvertebrate communities to progressive forest harvesting: Influences of harvest intensity, stream size and riparian buffers. Forest Ecology and Management 260: 1804-1815. doi:10.1016/j.foreco.2010.08.025.

39. Gravelle JA, Link TE, Broglio JR, Braatne JH (2009) Effects of Timber Harvest on Aquatic Macroinvertebrate Community Composition in a Northern Idaho. Watershed - Forest Science 55: 352-366.

40. Hotta N, Kayama T, Suzuki M (2007) Analysis of suspended sediment yields after low impact forest harvesting. Hydrological Processes 21: 3565-3575. doi:10.1002/hyp.6583.

41. Kreutzweiser DR, Capell SS, Good KP (2005) Macroinvertebrate community responses to selection logging in riparian and upland areas of headwater catchments in a northern hardwood forest. Journal of the North American Benthological Society 24: 208-222. doi:10.1899/0887-3593(2005)024.

42. Wang X, Burns DA, Yanai RD, Briggs RD, Germain RH (2006) Changes in stream chemistry and nutrient export following a partial harvest in the Catskill Mountains, New York, USA. Forest Ecology and Management 223: 103-112. doi:10.1016/j.foreco.2005.10.060.

43. Hutchens JJ Jr., Batzer DP, Reese E (2003) Bioassessment of silvicultural impacts in streams and wetlands of the eastern United States. Water Air, Soil Pollut 00: 1-17. doi: 10.1023/b:wafo.0000012827.95431.b8

44. Macdonald JS, Beaudry PG, MacIsaac EA, Herunter HE (2003) The effects of forest harvesting and best management practices on streamflow and suspended sediment concentrations during snowmelt in headwater streams in sub-boreal forests of British Columbia, Canada. Canadian Journal Forest Research-Revue Canadienne de Recherche Forestiere 33: 1397-1407. doi:10.1139/x03-110.

45. Martin CW, Hornbeck JW (2000) Impacts of intensive harvesting on hydrology and nutrient dynamics of northern hardwood forests. Canadian Journal of Fisheries and Aquatic Sciences 57: 19-29. doi:10.1139/f00-106.

46. Barbour MT, Gerritsen J, Snyder BD, Stribling JB (1999) Chapter 7.1 single habitat approach: 1-meter kick net. Rapid bioassessment

protocols for use in streams and wadeable rivers: periphyton, benthic macroinvertebrates and fish - Second Edition EPA 841-B-99-002. Washington DC: United States Environmental Protection Agency, Office of Water.

47. Western Regional Climate Center (2012) Manzanita Lake Period of Record General Climate Summary - Precipitation. Available: . http://www.wrcc.dri.edu/cgi-bin/cliMAIN.pl?ca5311. Accessed 4 July 2013.
48. PRISM Climate Group (2010) PRISM Data Explorer. Oregon State University. Available: . http://prismmap.nacse.org/nn/index.phtml. Accessed 24 February 2012.
49. Mosley MP, McKercher AI (1993) Streamflow. In: DR Maidment. Handbook of hydrology. New York: McGraw-Hill. pp. 8.1-8.39.
50. Eaton AD, Clesceri LS, Rice EW, Greenberg AE, Franson MAH (2005) Standard methods for the examination of water and wastewater: centennial edition. Washington DC: American Public Health Association. 1368 pp.
51. Flosi G, Downie S, Hopelain J, Bird M, Coey R, et al. (1998) California salmonid stream habitat restoration handbook: Appendix M, specialty instruments. California Department of Fish and Game. Available: http://www.dfg.ca.gov/fish/resources/habitatmanual.asp. Accessed 25 May 2012.
52. Platts WS, Armour C, Booth GD, Bryant M, Bufford JL et al. (1987) Methods for evaluating riparian habitats with applications to management. USDA Forest Service General Technical Report INT-221. Available: .http://www.fort.usgs.gov/Products/Publications/3581/3581.pdf. Accessed 17 Febuary 2012.
53. Moulton SR II, Carter JL, Grotheer SA, Cufney TF, Short TM (2000) Methods of analysis by the U.S. Geological Survey National Water Quality Laboratory - processing, taxonomy, and quality control of benthic macroinvertebrate samples. United States Geological Survey Open-File Report 00-212. Available: . http://nwql.usgs.gov/OFR-00-212.shtml. Accessed 17 Febuary 2012.
54. Cuffney TF, Gurtz ME, Meador MR (1993) Methods for collecting benthic invertebrate samples as part of the National Water-Quality Assessment Program. United States Geological Survey Open-File Report 93-406. Available: .http://nc.water.usgs.gov/reports/abstracts/ofr93406.html. Accessed 17 Febuary 2012.

55. Fowler WB, Lopushinsky W (1989) An economical, digital meter for gypsum soil-moisture blocks. Soil Science Society Of America Journal 53: 302-305. doi:10.2136/sssaj1989.03615995005300010057x.

56. Pinheiro JC, Bates DM (2000) Mixed-Effects Models in S and S-PLUS. New York: Springer.

57. Rabe-Hesketh S, Skrondal A (2008) Multilevel and Longitudinal Modeling Using Stata. College Station, TX: Stata Press.

58. Roche LM, Allen-Diaz B, Eastburn DJ, Tate KW (2012) Cattle grazing and Yosemite toad (Bufo canorus Camp) breeding habitat in Sierra Nevada meadows. Rangeland Ecology and Management 65: 56-65. doi:10.2111/REM-D-11-00092.1.

59. StataCorp (2009) Stata/SE 11.1 for Windows. College Station, TX: StataCorp LP.

60. Fenn ME, Baron JS, Allen EB, Rueth HM, Nydick KR et al. (2003) Ecological effects of nitrogen deposition in the western United States. Journal of Biosciences 53: 404-420. Available online at: doi:10.1641/0006-3568(2003)053[0404:EEONDI]2.0.CO;2

61. Nijboer RC, Verdonschot PFM (2004) Variable selection for modelling effects of eutrophication on stream and river ecosystems. Ecological Modelling 177: 17-39. doi:10.1016/j.ecolmodel.2003.12.050.

62. Sickman JO, Melack JM, Clow DW (2003) Evidence for nutrient enrichment of high-elevation lakes in the Sierra Nevada, California. Limnology and Oceanography 48: 1885-1892. doi:10.4319/lo.2003.48.5.1885.

63. Wood PJ, Armitage PD (1997) Biological effects of fine sediment in the lotic environment. Environ Manage 21: 203-217. doi:10.1007/s002679900019. PubMed:9008071.

64. Jewett K, Daugharty D, Krause HH, Arp PA (1995) Watershed responses to clear-cutting: Effects on soil solutions and stream water discharge in central New Brunswick. Canadian Journal of Soil Science 75: 475-490. doi:10.4141/cjss95-069.

65. Swank WT, Vose JM, Elliott KJ (2001) Long-term hydrologic and water quality responses following commercial clearcutting of mixed hardwoods on a southern Appalachian catchment. Forest

Ecology and Management 143: 163-178. doi:10.1016/S0378-1127(00)00515-6.

66. Kreutzweiser D, Capell S, Good K, Holmes S (2009) Sediment deposition in streams adjacent to upland clearcuts and partially harvested riparian buffers in boreal forest catchments. Forest Ecology and Management 258: 1578-1585. doi:10.1016/j.foreco.2009.07.005.
67. Binkley D, Ice GG, Kaye J, Williams CA (2004) Nitrogen and phosphorus concentrations in forest streams of the United States. J Am Water Resour As 40: 1277-1291. doi:10.1111/j.1752-1688.2004.tb01586.x.
68. Dubrovsky NM, Burow KR, Clark GM, Gronberg JM, et al H. (2010) The quality of our nation's waters - nutrients in the nation's streams and groundwater, 1992–2004. United States Geological Survey Circular 1350.
69. Powlson DS, Addiscott TM, Benjamin N, Cassman KG, de Kok TM et al. (2008) When does nitrate become a risk for humans? J Environ Qual 37: 291-295. doi:10.2134/jeq2007.0177. PubMed: 18268290.
70. Johnson DW, Susfalk RB, Dahlgren RA (1997) Nutrient fluxes in forests of the eastern Sierra Nevada mountains, United States of America. Global Biogeochemical Cycles 11: 673-681. doi:10.1029/97GB01750.
71. Sickman JO, Melack JM, Stoddard JL (2002) Regional analysis of inorganic nitrogen yield and retention in high-elevation ecosystems of the Sierra Nevada and Rocky Mountains. Biogeochemistry 57: 341-374. doi:10.1023/A:1016564816701.
72. Roche LM, Kromschroeder L, Atwill ER, Dahlgren RA, Tate KW (2013) Water Quality Conditions Associated with Cattle Grazing and Recreation on National Forest Lands. PLOS ONE 8: e68127. PubMed: 23826370.
73. Gomi T, Moore RD, Hassan MA (2005) Suspended sediment dynamics in small forest streams of the Pacific Northwest. Journal of the American Water Resources Association 41: 877-898. doi:10.1111/j.1752-1688.2005.tb03775.x.
74. Hassan MA, Church M, Lisle TE, Brardinoni F, Benda L et al. (2005) Sediment transport and channel morphology of small, forested

streams. Journal of the American Water Resources Association 41: 853-876. doi:10.1111/j.1752-1688.2005.tb03774.x.

75. Cooper AB, Thomsen CE (1988) Nitrogen and phosphorus in streamwaters from adjacent pasture, pine, and native forest catchments. New Zealand Journal of Marine and Freshwater Research 22: 279-291. doi:10.1080/00288330.1988.9516300.

76. Walker TW, Syers JK (1976) Fate of phosphorus during pedogenesis. Geoderma 15: 1-19. doi:10.1016/0016-7061(76)90066-5.

77. Elser JJ, Bracken MES, Cleland EE, Gruner DS, Harpole WS et al. (2007) Global analysis of nitrogen and phosphorus limitation of primary producers in freshwater, marine and terrestrial ecosystems. Ecol Lett 10: 1135-1142. doi:10.1111/j.1461-0248.2007.01113.x. PubMed: 17922835.

78. Kiffney PM, Richardson JS, Bull JP (2003) Responses of periphyton and insects to experimental manipulation of riparian buffer width along forest streams. Journal of Applied Ecology 40: 1060-1076. doi:10.1111/j.1365-2664.2003.00855.x.

79. Kiffney PM, Richardson JS, Bull JP (2004) Establishing light as a causal mechanism structuring stream communities in response to experimental manipulation of riparian buffer width. Journal of the North American Benthological Society 23: 542-555. doi:10.1899/0887-3593(2004)023.

80. Wallace JB, Eggert SL, Meyer JL, Webster JR (1997) Multiple trophic levels of a forest stream linked to terrestrial litter inputs. Science 277: 102-104. doi:10.1126/science.277.5322.102.

81. Chizinski CJ, Vondracek B, Blinn CR, Newman RM, Atuke DM et al. (2010) The influence of partial timber harvesting in riparian buffers on macroinvertebrate and fish communities in small streams in Minnesota, USA. Forest Ecology and Management 259: 1946-1958. doi:10.1016/j.foreco.2010.02.006.

82. Hemstad NA, Merten EC, Newman RM (2008) Effects of riparian forest thinning by two types of mechanical harvest on stream fish and habitat in northern Minnesota. Canadian Journal Forest Research-Revue Canadienne de Recherche Forestiere 38: 247-256. doi:10.1139/X07-157.

83. Hubbart JA, Link TE, Gravelle JA, Elliot WJ (2007) Timber harvest impacts on water yield in the continental/maritime hydroclimatic region of the United States. Forest Science 53: 169-180.

84. Beschta RL, Bilby RE, Brown GW, Holtby LB, Hofstra TD (1987) Stream temperature and aquatic habitat: fisheries and forestry interactions. In: EO SaloT.W. Cundy. Streamside Management: Forestry and Fishery Interactions, Contribution 57. Seattle, WA: Univ. of Wash., Inst. of Forest Res. pp. 191-232.
85. Moore RD, Spittlehouse DL, Story A (2005) Riparian microclimate and stream temperature response to forest harvesting: A review. Journal of the American Water Resources Association 41: 813-834. doi:10.1111/j.1752-1688.2005.tb04465.x.
86. Bourque CPA, Pomeroy JH (2001) Effects of forest harvesting on summer stream temperatures in New Brunswick, Canada: an inter-catchment, multiple-year comparison. Hydrology and Earth System Sciences 5: 599-613. doi:10.5194/hess-5-599-2001. Available online at: doi:10.5194/hess-5-599-2001.
87. Wilkerson E, Hagan JM, Siegel D, Whitman AA (2006) The effectiveness of different buffer widths for protecting headwater stream temperature in Maine. Forest Science 52: 221-231.
88. Rempel RS, Carter JCH (1986) An experimental-study on the effect of elevated-temperature on the heterotrophic and autotrophic food resources of aquatic insects in a forested stream. Canadian Journal Zoology-Revue Canadienne de Zoologie 64: 2457-2466. doi:10.1139/z86-366.
89. Barton DR, Taylor WD, Biette RM (1985) Dimensions of riparian buffer strips required to maintain trout habitat in southern Ontario Canada streams. North American Journal of Fisheries Management 5: 364-378. doi:10.1577/1548-8659(1985)5.
90. Hokanson KEF, Kleiner CF, Thorslund TW (1977) Effects of constant temperatures and diel temperature-fluctuations on specific growth and mortality-rates and yield of juvenile rainbow-trout, salmo-gairdneri. Journal of the Fisheries Research Board of Canada 34: 639-648. doi:10.1139/f77-100.
91. Carmona-Catot G, Moyle PB, Aparicio E, Crain PK, Thompson LC et al. (2010) Brook trout removal as a conservation tool to restore Eagle Lake rainbow trout. North American Journal of Fisheries Management 30: 1315-1323. doi:10.1577/M10-077.1.
92. Beche LA, McElravy EP, Resh VH (2006) Long-term seasonal variation in the biological traits of benthic-macroinvertebrates in

two Mediterranean-climate streams in California, USA. Freshwater Biology 51: 56-75. doi:10.1111/j.1365-2427.2005.01473.x.

93. Beche LA, Resh VH (2007) Short-term climatic trends affect the temporal variability of macroinvertebrates in California 'Mediterranean' streams. Freshwater Biology 52: 2317-2339. doi:10.1111/j.1365-2427.2007.01859.x.
94. Mykra H, Heino J, Muotka T (2008) Concordance of stream macroinvertebrate assemblage classifications: How general are patterns from single-year surveys? Biological Conservation 141: 1218-1223. doi:10.1016/j.biocon.2008.02.017.
95. Olden JD, Jensen OP, Vander Zanden MJ (2006) Implications of long-term dynamics of fish and zooplankton communities for among-lake comparisons. Canadian Journal of Fisheries and Aquatic Sciences 63: 1812-1821. doi:10.1139/f06-082.
96. Wiens JA (1981) Single-sample surveys of communities - are the revealed patterns real. American Naturalist 117: 90-98. doi:10.1086/283689.
97. Hawkins CP, Murphy ML, Anderson NH (1982) Effects of canopy, substrate composition, and gradient on the structure of macroinvertebrate communities in Cascade Range streams of Oregon. Ecology 63: 1840-1856. doi:10.2307/1940125.
98. Murphy ML, Hawkins CP, Anderson NH (1981) Effects of canopy modification and accumulated sediment on stream communities. Transactions of the American Fisheries Society 110: 469-478. doi:10.1577/1548-8659(1981)110.
99. Quinn JM, Boothroyd IKG, Smith BJ (2004) Riparian buffers mitigate effects of pine plantation logging on New Zealand streams 2. Invertebrate communities. Forest Ecology and Management 191: 129-146. doi:10.1016/j.foreco.2003.11.013.
100. Jost G, Weiler M, Gluns DR, Alila Y (2007) The influence of forest and topography on snow accumulation and melt at the watershed-scale. Journal of Hydrology 347: 101-115. doi:10.1016/j.jhydrol.2007.09.006.
101. Storck P, Kern T, Bolton S (1999) Measurement of differences in snow accumulation, melt, and micrometeorology due to forest harvesting. Northwest Science 73: 87-101.

102. Winkler RD, Spittlehouse DL, Golding DL (2005) Measured differences in snow accumulation and melt among clearcut, juvenile, and mature forests in southern British Columbia. Hydrological Processes 19: 51-62. doi:10.1002/hyp.5757.

103. Pomeroy JW, Gray DM, Hedstrom NR, Janowicz JR (2002) Prediction of seasonal snow accumulation in cold climate forests. Hydrological Processes 16: 3543-3558. doi:10.1002/hyp.1228.

104. Kaufmann MR (1985) Annual transpiration in sub-alpine forests - large differences among 4 tree species. Forest Ecology and Management 13: 235-246. doi:10.1016/0378-1127(85)90037-4.

105. Dunford EG, Niederhof CH (1944) Influence of aspen, young lodgepole pine, and open grassland types upon factors affecting water yield. Jour Forest 42: 673-677.

106. Pomeroy JW, Parviainen J, Hedstrom N, Gray DM (1998) Coupled modelling of forest snow interception and sublimation. Hydrological Processes 12: 2317-2337. doi:10.1002/(SICI)1099-1085(199812)12:15.

107. Jones BE, Lile DF, Tate KW (2011) Cattle selection for aspen and meadow vegetation: implications for restoration. Rangeland Ecology and Management 64: 625-632. doi:10.2111/REM-D-10-00089.1.

Chapter 7

Evidence of a High Density Population of Harvested Leopards in a Montane Environment

Julia N. Chase Grey[1,2*], Vivien T. Kent[1,3], Russell A. Hill[1,2]

[1]Department of Anthropology, Durham University, Durham, United Kingdom,

[2]Primate and Predator Project, Lajuma Research Centre, Louis Trichardt (Makhado) , South Africa,

[3]Durham Wildlife Trust, Houghton-le-Spring, United Kingdom

ABSTRACT

Populations of large carnivores can persist in mountainous environments following extensive land use change and the conversion of suitable habitat for agriculture and human habitation in lower lying areas of their range. The significance of these populations is poorly understood, however, and little attention has focussed on why certain mountainous areas can hold high densities of large carnivores and what the conservation implications of such populations might be. Here we use the leopard (*Panthera pardus*) population in the western Soutpansberg Mountains, South Africa, as a model system and show that montane habitats can support high numbers of leopards. Spatially explicit capture-recapture (SECR) analysis recorded the highest density of leopards reported outside of state-protected areas in sub-Saharan Africa. This density represents a temporally high local abundance of leopards and we explore the explanations for this alongside some of the potential conservation implications.

INTRODUCTION

Carnivore density and distribution is limited by a range of ecological and anthropogenic factors. One of the main constraints of carnivore density is prey availability and abundance, with carnivore densities being positively correlated with the density of their prey [1,2]. Large carnivores also require sufficient suitable habitat to provide cover for hunting and females require denning sites for rearing their young [3,4]. Anthropogenic factors limiting carnivore density include habitat loss [5], decline of prey densities due to hunting by humans [6], human persecution [7] and unsustainable harvest levels [8].

Widespread conversion of habitat into land for cultivation and human habitation has resulted in extensive habitat loss for many large carnivores. As a consequence, populations of felids, such as leopards (*Panthera pardus*) and pumas (*Puma concolor*), occur in mountainous areas that are less accessible to humans, where they persist after extirpation from lower lying altitudes of their ranges [9-15]. The conservation significance of these populations is poorly understood.

The leopard is one of the most widely distributed of the Felidae and inhabits a broad range of different habitats [9,16]. Their large geographic range is partially explained by their highly adaptable feeding behaviour which allows them to live wherever there is a sufficient prey base and hunting cover [17,18]. Nevertheless, global population numbers have declined over the last 100 years [19] and leopards have disappeared from 36.7% of their historical range in Africa [5]. The species was recently reclassified by the IUCN (International Union for Conservation of Nature) as "Near Threatened" and some of the leopard's most dramatic range loss has occurred in South Africa [9,20]. Decline of leopard habitat and populations can be due to habitat loss and fragmentation [5], poorly managed harvest quotas [8,21] and persecution [16,20]. Furthermore, there is a lack of widespread scientific input in quota setting for legal trophy hunting in many countries and few data exist on leopard numbers or metapopulation dynamics in many areas in which they are hunted [22]. Here we address the need for more data on leopard populations subject to both legal and illegal hunting and persecution through an examination of the population status of the leopard in the western Soutpansberg Mountains, South Africa. In identifying the existence of a temporally high population of leopards in the mountains, we discuss the factors contributing to this high density and the implications for conservation management and metapopulation dynamics.

MATERIALS AND METHODS

Ethics Statement

All fieldwork was approved by the Life Sciences Ethical Review Process Committee at Durham University, UK, and the Department of Anthropology Ethics Committee, with the ethics guidelines of the Association of Social Anthropologists of the UK and Commonwealth adhered to when interviewing landowners. All work was conducted with approved permits from the Limpopo Department of Economic Development, Environment & Tourism, South Africa.

Study Area

The study was conducted in the western part of the afro-montane forests of the Soutpansberg Mountains, Limpopo Province, South Africa (Figure 1). The mountains cover approximately 600km^2 and range in height from 250m above sea level to the highest peak Letjume (1748m) at the western extremity [23]. Temperatures vary in the wet season (December-February) from 16-40°C and in the dry season (May-August) between 12 and 22°C [24]. The western Soutpansberg is part of the Vhembe Biosphere Reserve, recognised as a hotspot of South African biodiversity and endemism [25].

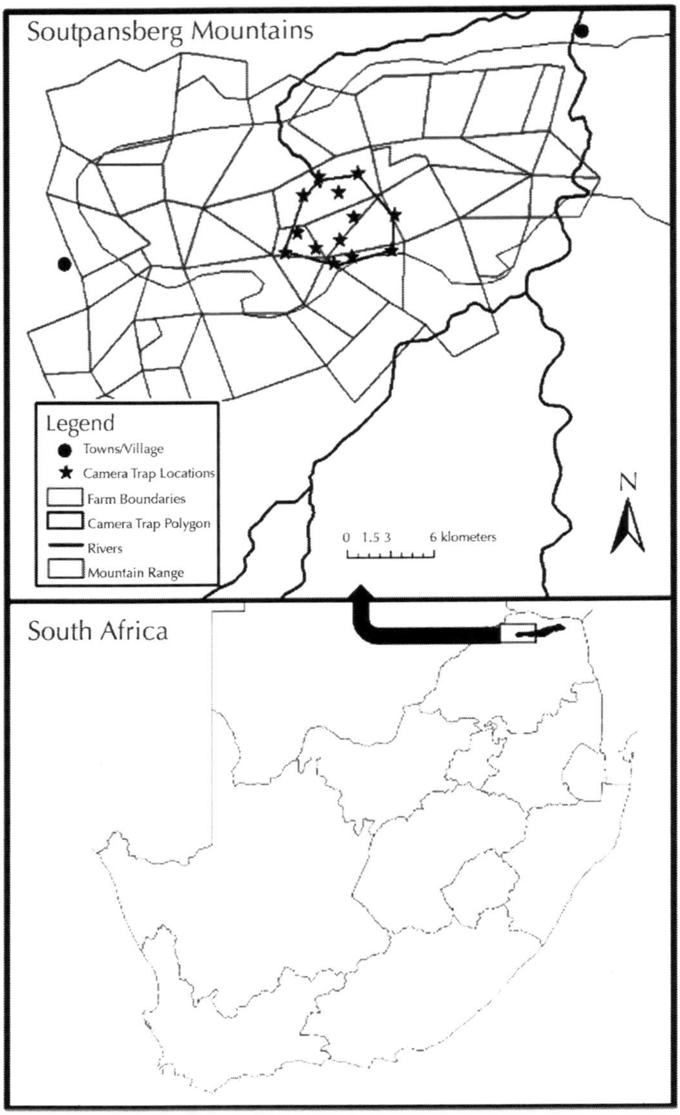

Figure 1: Camera trap study area within the western Soutpansberg Mountains. Lower map shows the location of the mountain range in Limpopo Province, South Africa.

Land use in the western Soutpansberg consists of a patchwork of private cattle and game farms, ecotourism properties, conservancies and communal farm land. Uncontrolled hunting during the 19th century and the destruction of habitat from cattle farming led to the extinction of mammals such as the African elephant (*Loxodonta africana*) and the black rhinoceros (*Dicero bicornis*) [26,27]. The only large carnivore species that remain resident in the western part of the mountain range are leopards, brown hyaena (*Parahyaena brunnea*) and spotted hyaena (*Crocuta crocuta*) [28].

Sampling Design and Field Methods

Leopard population density was determined through the application of a Bayesian spatially explicit capture-recapture (SECR) model to data acquired through camera trapping, a population sampling technique that allows researchers to estimate population densities of species that can be individually recognised from their natural markings [29,30]. Since the development of camera trapping, the methodology has been used extensively to obtain density estimates for a wide range of carnivores [31-38].

A camera trapping survey was carried out from March to May 2008. Following a two month scat and track survey, 13 paired camera trap stations were set up over an area of 31km^2 along roads and trails known to be frequented by leopards [39]. Camera stations were set up at a minimum distance of 1.7 km and a maximum distance of 3.5 km in order to ensure that all individuals in the study area had a probability of capture and that there were no gaps large enough to contain the home range of an individual leopard [39]. An area of 9 km^2 was used as the smallest recorded home range size for an adult female leopard in a forested habitat [40] and so, to ensure even camera coverage, at least 2 camera trap pairs were positioned within an area of this size [31].

Each camera trap station consisted of a pair of Cuddeback® Expert Digital Scouting Cameras (Non Typical Inc, USA) placed facing each other on either side of a road or trail. The cameras in each pair were placed a maximum of 6 metres apart to ensure that photographs of entire leopards were obtained and at a height of 40cm corresponding approximately to the shoulder height of an adult leopard [31]. Cameras were fixed to trees or set on bamboo stakes and fitted with roofs to protect them from sunlight and adverse weather conditions. The cameras were set to run continuously and were programmed to the smallest delay available for the models (one minute). The mountainous topography of the site made it impractical to move cameras to new positions during the survey and so cameras remained in fixed positions.

In order to meet the assumption of population closure, such that the sampling period is short enough so that no births, deaths, or emigration/immigration occur, the survey lasted for 63 days (9 weeks) [39]. To prevent battery failure, cameras were checked every two weeks with images downloaded from memory cards onto a laptop. Once images were saved, the date, time and location of each photograph was noted. All cameras were functional for the full duration of the study.

Spatially Explicit Capture-Recapture Camera Trapping

Each camera trapping night was used as one trapping occasion. The sex of each photographed leopard was established via the presence of external genitalia and relative body size and individuals were identified via their unique spot patterns. Due to the difficulty of correctly assigning leopards to narrow age categories from photographs [41], individuals were put in two broad age classes, juvenile (less than 2 years old) or adults (above 2 years old) in order to prevent confounding the data [41]. Age classification was conducted by multiple researchers via the examination of three diagnostic characteristics: extent of facial scarring, ear condition and dewlap size for males [41].

A Bayesian SECR model was used to estimate leopard density from photographs of individual leopards and capture locations. SECR provides a more accurate measurement for the effective survey area than solely using capture-recapture analysis as it uses the locations where each animal is detected to fit a spatial model of the detection process [42,43]. SECR obtains estimates of population density unbiased by edge effects, incomplete detection and heterogeneous capture probabilities and eliminates the need for an ad hoc estimation of the sampling area [44-47]. SECR analysis was conducted using SPACECAP version 1.0.6 [48] in R version 2.15.2 [49].

SPACECAP requires three data input files: 1) animal capture details (information on animal identification, trap location and sampling occasion), 2) trap deployment details (spatial location, dates when specific traps were active and sampling occasion) and 3) state-space details. In SPACECAP analyses, the surveyed area containing the camera trap array is combined with an extended area surrounding it, known as the "state-space" of the underlying point process, S, which is represented by a large number of equally spaced points in the form of a very fine mesh. These points are visualized as representing all possible potential activity centres (or home range centres) of all the animals in the population being surveyed [48]. The state space area equals the outer camera trap rectangle surrounded by a large buffer area. 'S' was created using ArcGIS 9.3 (ESRI, Redlands, USA) and was made up of UTM coordinates of potential home range centres created as equally spaced points and an associated column indicating habitat suitability at each point. A buffer distance, which is sufficiently large to ensure that no individual animal outside of the buffered region has any probability of being photo-captured by the camera traps in the array during the survey, was added to the rectangle encompassing the trap array. Then, using ArcGIS 9.3, numerous equally spaced points representing home range centres were generated for this extended area [48]. For the state-space file (the buffered area including the camera polygon) the mesh size was set at 1 km. Habitat not thought suitable for leopards (arable areas) was removed from the analysis. The areas removed were large, flat crop circles that provided no hunting cover for leopards. Interviews with local farmers confirmed that leopards were not seen in these areas. The model was run with buffers of 10, 15, 20 and 25 km, both with and without suitable habitat removed in order to examine 1) the effect of the habitat mask on the resulting density estimate [50] and 2) at what point the density estimate stabilised.

The following model definitions were used for the analysis: trap response present, spatial capture–recapture, half-normal detection function and Bernoulli's encounter model. The 'trap response present' option runs the behavioural response option (equivalent to Model "Mb"). This model implements a local or "trap-specific" behavioural response under which the probability of encounter in a trap increases subsequent to initial capture in that trap [48]. The number of Markov-Chain Monte Carlo (MCMC) iterations was set at 100,000 with a burnin period of 20,000 iterations and a thinning rate of 1. The analysis was run with a data augmentation value of 375 (37.5 times the number of animals identified in the survey) in order to achieve chain convergence for all parameters [51]. Chain convergence was assessed via examination of z score values produced by the Geweke diagnostic statistic in SPACECAP. Z scores greater than 1.6 imply that the MCMC analysis has not been run long enough and chains have not converged. A Bayesian P-value is produced by SPACECAP allowing for assessment of the adequacy of the model, values close to 0 or 1 imply that the model is inadequate [48].

Prey Abundance

In order to examine the abundance of prey species in the western Soutpansberg Mountains, an index of species abundance and a relative abundance index were calculated from camera trap photographs taken during the survey. Relative abundance indices from camera trap surveys have been shown to be directly related to independently derived density estimates of these species [32].

To calculate prey species abundance, each photograph of an animal was identified to species level and the time and date recorded, with the photo then classified as an independent or dependent event. An independent event was defined as consecutive photos of different species or consecutive photographs of individuals of the same species taken more than 1 hour apart [52]. A RAI was then calculated from the camera trapping data. The RAI equalled the number of days required to obtain a photograph of each species. This index measured effort and was expected to decrease as density increased with a score of 0 indicating that the species was captured on the day the camera trap survey began, required low capture effort and existed at a high density on the study site [53]. Species with higher RAI scores took a greater capture effort to obtain a photograph and therefore signified a lower density.

RESULTS

Leopard Density

103 photographs (53 left flanks and 50 right flanks) of 10 individual adult leopards (7 females and 3 males) were obtained over 819 trap nights, alongside 31 photographs (15 left flank and 16 right flank) of 4 juvenile individuals (2 males and 2 females) (Table 1). Capture histories of individually identified leopards were used to calculate capture frequencies. The number of captures and recaptures ranged from 1 to 19. Only adult leopards were used to calculate leopard densities as cubs of solitary felids (age <1 year) have been found to have low capture probabilities and sub-adults may be transient individuals and therefore not part of the resident population [54]. Spatially explicit capture recapture via SPACECAP provided a density estimate of 25 adults per 100km^2 (SD 8.54) with a buffer of 10km. The density estimates then stabilised between 10.85 (SD 3.68) and 10.22 (SD 3.00) for buffers between 15km and 25km (Figure 2). The removal of unsuitable leopard habitat had little effect on the final result. Since a buffer of 20km is larger than the home range of a female leopard collared in the western Soutpansberg (14km^2) [55], this buffer was selected to ensure that the buffer area was large enough such that no individual animal outside of the buffered region had any probability of being photo-captured by the camera traps during the survey. With a 20km buffer the mean density estimate was 10.73 adult leopards per 100 km^2 with a standard deviation of 3.32 (Table 2). Without unsuitable habitat removed from the state space area, the density was 10.41 adult leopards per 100 km^2 with a standard deviation of 3.13 suggesting that the unsuitable habitat did not significantly influence the density estimates. The density estimate obtained with a 20km buffer and unsuitable habitat removed is thus robust and a Bayesian P value of 0.60 confirmed the adequacy of the model with Z scores from the Geweke diagnostic confirming that chain convergence had been achieved for all parameters (Table 2). The encounter probabilities (posterior means) for individuals pre- and post-initial encounter were $p_1 = 0.0286$ and $p_2 = 0.649$ indicating a positive trap response. This is unlikely to reflect "trap-happiness", however, but instead indicates non-independence among encounters due to leopards favouring certain trails in moving through their territories [47].

Table 1: Individuals and age sex classes identified from camera images in the western Soutpansberg Mountains, South Africa

Leopard ID Code	Total captures	Number of capture locations
AF1	6	3
AF2	11	4
AF3	4	1
AF4	2	1
AF5	4	2
AF6	1	1
AF7	1	1
AM1	2	1
AM2	10	6
AM3	14	3
JF1	5	1
JF2	4	1
JM1	2	2
JM2	3	1

AF (adult female), AM (adult male), JF (juvenile female) and JM (juvenile male).

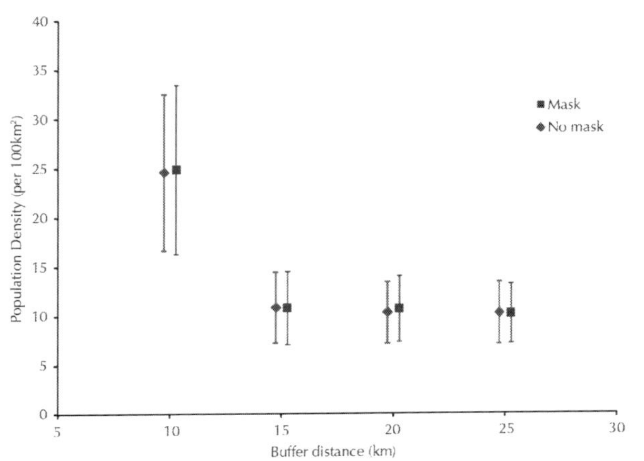

Figure 2: Graph showing the effect of buffer increase and use of a habitat mask on the SPACECAP density estimate (mean and standard deviation).

Table 2: Posterior summary statistics and z scores from the SECR model fitted to the leopard camera trapping data in the western Soutpansberg Mountains, South Africa

Parameter	Mean	SD	95% lower HPD level	95% upper HPD level	Z scores
D	10.73	3.32	4.68	17.33	
λ_0	0.029	0.0089	0.013	0.047	-0.86
σ	1.69	0.26	1.25	2.23	-0.71
$b1$	1.10	0.32	0.46	1.72	-0.22
Ψ	0.54	0.17	0.23	0.87	0.22
Ns	206.16	63.81	89.00	332.00	0.22

Only adult leopards were included. D is density/100 km^2; $\lambda_0$is the expected encounter rate; σ= sqrt(1/b^2) is the scale parameter of a bivariate normal encounter function and may also be viewed as a range parameter of an animal; $b1$ is the regression coefficient measuring the behavioural response; Ψ is the ratio of the number of animals actually present within the state space to the maximum allowable number; Ns is Nsuper and equals the population size for the state space.

Prey Abundance

Table 3 shows the capture frequencies of prey species expressed as number of photos per 100 camera days and the results of the relative abundance index. Bushbuck (*Tragelaphus scriptus*) were the most frequently captured species in camera trap photographs with a capture frequency of 1.56 per 100 traps nights. They were also one of the prey species that took the least measured effort (0 days) to obtain a photograph. Other small to medium sized antelopes known to be preferred leopard prey [3] included impala (*Aepyceros melampus*), red (*Cephalophus natalensis*) and common duiker (*Sylvicapra grimmia*). These species were captured at lower frequencies of 0.31 (impala), 0.17 (red duiker) and 0.12 per 100 trap nights (common duiker). Capture effort for impala and red duiker was relatively low compared to other species (RAI = 3), although the RAI value was higher for common duiker.

Table 3: Capture frequencies expressed as number of independent photos per 100 camera days and a relative abundance index calculated from capture effort for 18 species in the western Soutpansberg Mountains, South Africa

Species	Capture frequency (per 100 camera days)	Relative abundance index (capture effort)
Bushbuck	1.56	0
Porcupine	1.09	3
Baboon	1.08	0
Kudu	0.5	5
Giraffe	0.37	3
Impala	0.31	3
Lesser spotted genet	0.31	9
Warthog	0.31	2
Civet	0.3	5
Helmeted guinea fowl	0.29	0
Aardvark	0.17	2
Red duiker	0.17	3
Cattle	0.14	6
Sable	0.13	6
Common duiker	0.12	7
Donkey	0.12	6
Bush pig	0.1	19
Nyala	0.09	7

DISCUSSION

The results of this study provide evidence of a high density estimate of leopards in the montane habitat of the western Soutpansberg Mountains (10.7 per 100 km^2). This density estimate is the highest adult leopard population density recorded outside a state-protected area in Africa (Table 4). Areas of greater leopard density have been reported in the Sabie river area of southern Kruger National Park, South Africa (30.3 leopards per 100km^2) [56], Phinda Private Game Reserve [7] (11.25 per 100km^2), the N'wanetsi concession, Kruger National Park (12.7 per 100km^2) [57] and Ivindo National Park in Gabon (12.1 per 100km^2)

[58], but these all occur in national parks. Our reported leopard density greatly exceeds densities reported in other mountainous areas such as in the Cederberg and Waterberg Mountains, South Africa, where density ranges from 0.62 to 3 per 100km² [12,14]. It is important to note, however, that our study only represents a point density estimate and longer term camera trapping surveys need to be undertaken in order to examine population fluctuations and the effects of factors such as seasonality on leopard population numbers. Nevertheless, our results suggest a significant leopard population within the Soutpansberg Mountains.

Table 4: Leopard densities reported for previous studies across sub-Saharan Africa

Location	Density (per 100km²)	State-protected	Habitat type	Reference
Kruger National Park, South Africa	30.3	Y	Riverine forest	[56]
Ivindo National Park, Gabon	12.1	Y	Forest	[58]
N'wanetsi concession, Kruger National Park	12.7	Y	Savannah woodland	[57]
Phinda Private Game Reserve	11.25	Y	Woodland and grassland	[7]
Western Soutpansberg, Limpopo, South Africa	10.7	N	Montane woodland	This study
Tai National Park, Ivory Coast	8.7	Y	Forest	[76]
Tsavo National Park, Kenya	7.7	Y	Montane bushveld	[77]
Ranches, Laikipia District, Kenya	5.5 - 8.5	N	Savannah and woodland	[78]
Serengeti National Park, Tanzania	4.7	Y	Forest	[79]
Serengeti National Park, Tanzania	3.8	Y	Forest	[80]
North-central farmland, Namibia	3.6	N	Shrub and woodland	[81]
Kruger National Park, South Africa	3.5	Y	Savannah	[56]
Waterburg, South Africa	3.0	N	Montane savannah	[12]

Logging Concession, Gabon	2.7	N	Forest	[58]
Cederberg Wilderness Area, South Africa	2.0	Y	Montane fynbos	[14]
Kaudom Game Reserve, Namibia	1.5	N	Savannah	[82]
Waterberg Plateau Park, Namibia	1.0	Y	Savannah and woodland	[81]
Cederberg Wilderness Area, South Africa	0.62	Y	Montane karoo	[14]
Kalahari Gemsbok National Park South Africa	0.6	Y	Savannah	[83]

There are a number of possible reasons for the high density recorded in the Soutpansberg. Firstly, it may represent a temporally local high abundance of leopards. Leopard population numbers can fluctuate significantly over time and a study in the Karonge Game Reserve, South Africa, found that leopard density changed from 20 per 100km^2 to 5 per 100km^2 over a five year period [59]. Population numbers may be affected by short-term changes in the environment and in population dynamics which can cause variance in breeding success [60]. The possibility exists, therefore, that the densities recorded here may not be stable in the long term and further work examining the population dynamics of these leopards is required.

Carnivore abundance is also limited by the biomass of its prey, and variation in leopard density is partly due to differences in biomass and abundance of prey species [1,2]. Leopards preferentially feed upon animals between 10 and 40 kg that live in small herds, occupy dense habitat and afford minimal risk of injury during capture such as bushbuck, common and red duikers [3,55]. Bushbuck, the most frequently taken prey item by relative frequency in leopard diets in the western Soutpansberg [55], was both the most commonly captured species in camera trap photographs and one of the prey species that took the least measured effort to obtain a photograph. This suggests that the survey area holds a high abundance of bushbuck, a preferred leopard prey species, which may account for the high density of leopards on the study site. Although the RAI does provide on information on prey abundance, it was not possible to collect independently derived data on leopard prey species due to the highly mountainous terrain which made conducting line transects impractical [61]. Nevertheless, known preferred prey species such as bushbuck and other small to medium antelope are well represented in our data. Since the RAI provides an indication of the abundance of these species at the survey site it suggests a high prey base of preferred species within the Soutpansberg Mountains.

A modelling study to estimate the extent of suitable leopard habitat in South Africa found surface ruggedness to be one of the highest contributing variables underlying the most parsimonious habitat suitability model [9]. Studies on cougars [62], jaguars (*Panthera onca*) [63] and leopards [64] have similarly found rugged topography to be important habitat for large predators. Our study area encompassed the highest peak (Letjume) in the western Soutpansberg Mountains [23] with significant on-site altitudinal variation [65,66]. As a consequence the region may constitute prime leopard habitat in terms of topography. Montane areas are difficult for humans to access, have lower human activity than less rugged terrain and thus offer isolated habitat with less direct competition for space and lower anthropogenic persecution than in lower lying areas [9]. Little or no cattle or game farming occurs in the highest elevations in the study area [55] further reducing the anthropogenic pressure and increasing the suitability of the habitat for leopards.

Finally, land use within the Western Soutpansberg appears compatible with leopard conservation. Land use consists of private cattle and game farms, ecotourism properties, conservancies and communal farm land [55]. The camera trapping survey covered seven properties and an attitudinal survey of landowner perceptions towards leopards found that the majority of landowners (71%) who were engaged in consumptive and non-consumptive use of leopards (ecotourism and game farms) reported positive attitudes towards leopards [55]. Previous studies have also found landowners belonging to conservancies and eco-tourism operators to hold more positive attitudes to carnivores than those involved with livestock farming [67]. Although landowners engaged in cattle and small-stock farming have been found to be responsible for the bulk of leopard persecution in the area [55], as well as being one of the main causes of habitat fragmentation and persecution of leopards due to human wildlife conflict [9], such properties were at a low density in the upper reaches of the mountains [55]. As a consequence, land use in the study area may be more compatible with leopard conservation despite the potential influence of edge effects caused by the mosaic of land use types [7].

Attractive Sink or Population Source?

High density populations are often assumed to be population sources [68], but this conclusion ignores the possibility that areas with high population numbers may be acting as attractive sinks. Attractive sinks contain high numbers of dispersing sub-adults that enter an area with favourable habitat and unoccupied territories made vacant due to factors such as overharvesting [69]. Large numbers of dispersers can artificially inflate population numbers making the area appear to be a population source [70]. This short term inflation of population numbers can mask an overall metapopulation decline as dispersers are drawn into the sink and are then affected by the high mortality rates acting there. Large numbers of sub-adult individuals within a population are thus indicative of high population turnover and suggest a region is acting as an attractive sink.

Fourteen individual leopards were identified in this study of which ten were adults and four juveniles (Table 1). Due to the difficulty in aging leopards from camera trap photographs [41], sub-adults [2-3 years] and males less than 7 years were not classified in separate age groups. Although the demographic information presented in Table 1 could suggest a stable population, data from long-term (5 years) unpublished camera trapping records from the same area indicates only one of the adult males (AM2) has been continually present. This may imply a high turnover of sub-adult males and males under 7 years and so suggests that the area is functioning as a potential attractive sink. However, longer term data on population density and dynamics and the categorisation of individuals into narrower age classes is required in order to confirm this hypothesis.

Methodological Considerations

The density estimate obtained via SECR (10.7 leopards per 100km^2) is high in comparison to leopard densities calculated from other mountainous areas (Table 4). It is important to recognise however, that certain factors in survey design and analysis may bias density estimates. Recent research has shown that population estimates in SECR can be positively biased by the use of small camera trapping polygons [71-73]. Tobler & Powell [73] suggested that camera trapping polygons should be at least the size of the largest home range of a male in order to prevent bias from small survey size. During the course of our study, no data were available on the home ranges of adult males. A subsequent GPS satellite telemetry study provided home range data for one adult female (16.3km² 95% kernel estimate) [55]. If no overlap occurred between females and an adult male home range overlapped with approximately four females [56], a male range might extend to 65km². Our camera polygon of 31km² was thus small compared to the recommendation of Tobler & Powell [73] and this may have positively biased the density estimate. Nevertheless, SECR models can produce unbiased density estimates even when the camera polygon size is about half the size of a home range of a male jaguar [73,74] and further work will be required to confirm the stability of our estimate.

A second consideration with our density estimate is the absence of sex covariates in the SECR analysis. Currently there is no facility in SPACECAP to use covariates, but sex and age can influence the movement and detection parameters, and therefore density [72,73]. Heterogeneous capture probabilities between males and females biases camera trapping data towards the sex with the greater encounter probability, in this case males [71], and this creates a negative bias in both traditional mark-recapture and spatial capture-recapture models leading to an underestimate of density [73]. One option to remove this bias is to estimate density separately for males and females, but this is only possible if sample sizes are adequate [71] which was not the case for our study (N=10 adult leopards). Nevertheless, since SECR models with sex covariates have shown that not accounting for sex could result in an underestimation of the true density estimate [72], future studies could improve the precision of density estimates through the inclusion of sex covariates.

Finally, research has shown that estimating the age of leopards based on photographs is difficult and can lead to misclassification of individuals to age groups [41], particularly for sub-adults (2-3 years of age). As a consequence, density estimates may be biased by the classification of some juveniles as adults, thus inflating the density estimate. Although based on our methods we are confident that all individuals were aged and sex correctly, incorrect age estimation remains an area of potential bias. If some juveniles were misclassified as adults, this could imply a population made up of a greater proportion of younger individuals, again suggesting that the survey site may be functioning as an attractive sink for immigrating individuals. Longer term data on population density and dynamics of known-age individuals are required to explore this further.

Conservation and Management Implications

Where leopard populations are subject to illegal hunting, poaching and snaring, or legal harvesting through trophy hunting, increased hunting pressure will lead to a high risk of long-term decline. Wildlife authorities therefore need to take leopard population dynamics into account when assigning trophy hunting permits. Mortality from trophy hunting is additive to other forms of anthropogenic mortality. A recent call has been made to increase trophy hunting permits for leopards in the western Soutpansberg by local hunting factions who see the area as being a high density source for leopard trophies [55].

The results of this study provide evidence of a high density estimate of leopards in the afro-montane forest habitat of the western Soutpansberg Mountains (10.7 per 100 km^2). This density estimate is the highest adult leopard population density recorded for a non-protected area in Africa (Table 4). Long term data are now needed on leopard densities over a larger area in the western Soutpansberg in order to confirm the conservation importance of this area. Analysis of these data must include the use of sex covariates, a survey area size that does not positively bias density results and the use of data on known-age individuals. Until these data are available calls to increase trophy hunting permits in the Soutpansberg should be resisted to ensure that the leopard population is not being over-harvested. An improvement in livestock husbandry practices in the study site is also required to mitigate human-wildlife conflict and illegal carnivore persecution and reduce the negative effects of anthropogenic persecution on leopard population dynamics [75].

CONCLUSIONS

The western Soutpansberg Mountains in South Africa are home to the highest density of leopards recorded outside a state-protected area in sub-Saharan Africa. As leopards have experienced a 36.7% range loss across Africa, and only 20% of South Africa now holds suitable habitat for leopards, this mountain population may be of conservation importance for the wider leopard metapopulation [5,12]. Research has shown that unprotected, mostly privately owned land is extremely important for South African leopard conservation and in Limpopo Province 95% of suitable leopard habitat is still situated outside protected areas [12]. Leopard conservation efforts should therefore be focused on private land such as the Soutpansberg Mountains in order to preserve the status of this large carnivore.

ACKNOWLEDGMENTS

We are grateful to Dr Ian Gaigher for permission to work at the Lajuma Research Centre and to Oldrich Van Schalwyk, Ian Gaigher and Lourens Swanepoel for discussion on Soutpansberg leopard population dynamics. We thank Arjun Gopalaswamy, Andrew Royle and James Hines for advice on SPACECAP. We are also grateful to Sian Waters and Rachel Sassoon for independent assessments of leopard age and sex. We are indebted to two anonymous referees for their thoughtful reviews that significantly improved the manuscript.

REFERENCES

1. Karanth KU, Nichols JD, Kumar NS, Link WA, Hines JE (2004) Tigers and their prey: predicting carnivore densities from prey abundance. Proc Natl Acad Sci U S A 101: 4854-4858. doi:10.1073/pnas.0306210101. PubMed: 15041746.
2. Carbone C, Gittleman JL (2002) A common rule for the scaling of carnivore density. Science 295: 2273-2276. doi:10.1126/science.1067994. PubMed: 11910114.
3. Hayward MW, Henschel P, O'Brien P, Hofmeyr M, Balme G et al. (2006) Prey preferences of the leopard (*Panthera pardus*). J Zool Lond 270: 298–313. doi: 10.1111/j.1469-7998.2006.00139.x
4. Mizutani F, Jewell PA (1998) Home-range and movement of leopards (*Pantherapardus*) on a livestock ranch in Kenya. J Zool Lond 244: 269-286. doi:10.1111/j.1469-7998.1998.tb00031.x.

5. Ray JC, Hunter L, Zigouris J (2005). Setting conservation and research priorities for larger African carnivores. WCS Working Paper No. 24 New York: Wildlife Conservation Society.
6. Henschel P, Hunter LTB, Coad L, Abernethy KA, Mühlenberg M (2011) Leopard prey choice in the Congo Basin rainforest suggests exploitative competition with human bushmeat hunters. J Zool Lond 285: 11–20. doi: 10.1111/j.1469-7998.2011.00826.x
7. Balme GA, Slowtow R, Hunter LTB (2010) Edge effects and the impact of non-protected areas in carnivore conservation: leopards in the Phinda-Mkhuze Complex, South Africa. Anim Conserv 13: 315-323. doi:10.1111/j.1469-1795.2009.00342.x.
8. Packer C, Brink H, Kissui BM, Maliti H, Kushnir H et al. (2011) Effects of trophy hunting on lion and leopard populations in Tanzania. Conserv Biol 25: 142-153. doi:10.1111/j.1523-1739.2010.01576.x. PubMed: 20825444.
9. Swanepoel LH, Lindsey P, Somers MJ, Van Hoven W, Dalrum F (2013) Extent and fragmentation of suitable leopard habitat in South Africa. Anim Conserv 16: 41-50. doi:10.1111/j.1469-1795.2012.00566.x.
10. Sweanor LL, Logan KA, Hornocker MG (2000) Cougar dispersal patterns, metapopulation dynamics, and conservation. Conserv Biol 14: 798-808. doi:10.1046/j.1523-1739.2000.99079.x.
11. Spalton JA, Al Hikmani HM, Willis D, Said Bait (2006) Critically Endangered Arabian leopards (*Panthera pardus nimr*) persist in the Jabal Samhan Nature Reserve, Oman. Oryx 40: 287-294. doi:10.1017/S0030605306000743.
12. Swanepoel L (2009) Ecology and conservation of leopards (*Panthera pardus*) on selected game ranches in the Waterberg region, Limpopo, South Africa. MSc Thesis, University of Pretoria, South Africa.
13. Cooley HS, Wielgus RB, Koehler GM, Maletzke BT (2009) Source populations in carnivore management: cougar demography and emigration in a lightly hunted population. Anim Conserv 12: 321–328. doi:10.1111/j.1469-1795.2009.00256.x.
14. Martins QE (2010) The ecology of the leopard (*Panthera pardus*) in the Cederberg Mountains. PhD thesis, University of Bristol, UK.

15. Ruth TK, Haroldson MA, Murphy KM, Buotte PC, Hornocker MG et al. (2011) Cougar survival and source-sink structure on Greater Yellowstone's Northern Range. J Wildlife Manage 75: 1381–1398. doi:10.1002/jwmg.190.
16. Balme GA, Slowtow R, Hunter LTB (2009) Impact of conservation interventions on the dynamics and persistence of a persecuted leopard (Panthera pardus). Population - Biol Conserv 142: 2681-2690. doi: 10.1016/j.biocon.2009.06.020
17. Hunter L, Balme G, Walker C, Pretorius K, Rosenberg K (2003) The landscape ecology of leopards in northern KwaZulu-Natal, South Africa: a preliminary project report. Ecol Journal 5: 24-30.
18. Hunter L (1999) The Quintessential Cat. Environ Wildl 7: 32-41.
19. Uphyrkina O, Johnson WE, Quigley H, Miquelle D, Marker Let al (2001) Phylogenetics, genome diversity and origin of modern leopard (*Panthera pardus*). Mol Ecol 10: 2617-2633. doi:10.1046/j.0962-1083.2001.01350.x. PubMed: 11883877.
20. Henschel P, Hunter L, Breitenmoser U, Purchase N, Packer C et al. (2008) *Pantherapardus*. In: IUCN Red List of Threatened Species. Version 2010.3. Gland: IUCN Species Survival Commission.
21. Packer C, Kosmala M, Colle HS, Brink H, Pintea AL et al. (2009) Sport hunting, predator control and conservation of large carnivores. PLOS ONE, 4: e5941. doi:10.1371/journal.pone.0005941. PubMed: 19536277.
22. Balme GA, Hunter LTB, Goodman P, Ferguson H, Craigie J et al. (2010) An adaptive management approach to trophy hunting of leopards (*Panthera pardus*): a case study from KwaZulu-Natal, South Africa. In: DW MacdonaldAJ Loveridge. Biology and conservation of wild felids. Oxford: Oxford University Press. pp. 341–352.
23. Mostert THC, Bredenkamp GJ, Klopper HL, Verwey C, Mostert RE et al. (2008) Major vegetation types of the Soutpansberg Conservancy and the Blouberg Nature Reserve, South Africa. Koedoe 50: 32-48. doi: 10.4102/koedoe.v50i1.125
24. Kabanda TA (2003) Climate. In: K. BergerJE CraffordI. GaigherMJ GaigherN. Hahn. A first synthesis of the environmental, biological and cultural assets of the Soutpansberg. Louis Trichardt: Leach Printers & Signs.

25. UNESCO (2009) UNESCO MAB Biosphere Reserves Directory. Available: .http://www.unesco.org/mabdb/br/brdir/directory/biores.asp?code=SAF+06&mode=al. Accessed March 1st 2010.
26. MacKenzie JM (1988) The empire of nature: Hunting, conservation and British imperialism. Manchester: Manchester University Press.
27. Hahn N (2006) Floristic diversity of the Soutpansberg, Limpopo Province, South Africa. PhD Thesis, University of Pretoria, South Africa.
28. Gaigher IG, Stuart CT (2003) Mammals. In: K. BergerJE CraffordI. GaigherMJ GaigherN. Hahn. A first synthesis of the environmental, biological and cultural assets of the Soutpansberg. Louis Trichardt: Leach Printers & Signs.
29. Karanth KU, Nichols JD (1998) Estimation of tiger densities in India using photographic captures and recaptures. Ecology 79: 2852-2862. Available online at:doi:10.1890/0012-9658(1998)079[2852:EOTDII]2.0.CO; 2.
30. Karanth KU, Chundawat RS, Nichols JD, Kumar NS (2004) Estimation of tiger densities in the tropical dry forests of Panna, Central India, using photographic capture-recapture sampling. Anim Conserv 7: 285-290. doi:10.1017/S1367943004001477.
31. Henschel P, Ray J (2003) Leopards in African Rainforests: Survey and Monitoring Techniques. New York: Wildlife Conservation Society.
32. O'Brien TG, Kinnaird MF, Wibisono HT (2003) Crouching tigers, hidden prey: Sumatran tiger and prey populations in a tropical forest landscape. Anim Conserv 6: 131–139. doi:10.1017/S1367943003003172.
33. Kawanishi K, Sunquist ME (2004) Conservation status of tigers in a primary rainforest of peninsular Malaysia. Biol Conserv 120: 329–344. doi:10.1016/j.biocon.2004.03.005.
34. Wegge P, Pokheral CP, Jnawali SR (2004) Effects of trapping effort and trap shyness on estimates of tiger abundance from camera trap studies. Anim Conserv 7: 251-256. doi:10.1017/S1367943004001441.
35. Jackson RM, Roe JD, Wangchuk R, Hunter DO (2006) Estimating snow leopard population abundance using photography and

capture-recapture techniques. Wildlife Soc Bull 34: 772-781. Available online at: doi:10.2193/0091-7648(2006)34[772:ESLPAU]2.0.CO; 2.

36. Balme GA, Hunter LTB, Slotow R (2009) Evaluating methods for counting cryptic carnivores. J Wildlife Manage 73: 433-441. doi:10.2193/2007-368.
37. Wang S, Macdonald D (2009) The use of camera traps for estimating tiger and leopard populations in the high altitude mountains of Bhutan. Biol Conserv 142: 606-613. doi:10.1016/j.biocon.2008.11.023.
38. Kent VT, Hill RA (2013) The importance of farmland for the conservation of the brown hyaena *Parahyaena brunnea*. Oryx 47: 431-440. doi:10.1017/S0030605312001007.
39. Karanth KU, Nichols JD (2002) Monitoring Tigers and Their Prey: A manual for Researchers, Managers and Conservationists in Tropical Asia. Bangalore: Centre for Wildlife Studies.
40. Grassman LI (1999) Ecology and behavior of the Indochinese leopard in Kaeng Karchan National Park, Thailand. Nat Hist Bull of Siam Soc 47: 77-93.
41. Balme GA, Hunter L, Braczkowski AR (2012) Applicability of Age-Based Hunting Regulations for African Leopards. PLOS ONE, 7: e35209. doi:10.1371/journal.pone.0035209. PubMed: 22493739.
42. Efford MG, Borchers DL, Byrom AE (2009) Density estimation by spatially explicit capture–recapture: likelihood-based methods. In: DL ThomsonEG CoochMJ Conroy. Modelling demographic processes in marked populations. New York: Springer. pp. 255–269.
43. Efford MG, Dawson DK, Borchers DL (2009) Population density estimated from locations of individuals on a passive detector array. Ecology 90: 2676-2682. doi:10.1890/08-1735.1. PubMed: 19886477.
44. Karanth KU, Nichols JD, Kumar NS, Hines JE (2006) Assessing tiger population dynamics using photographic capture–recapture sampling. Ecology 87: 2925–2937. Available online at: doi:10.1890/0012-9658(2006)87[2925:ATPDUP]2.0.CO;2. PubMed: 17168036.

45. Soisalo MK, Cavalcanti SMC (2006) Estimating the density of a jaguar population in the Brazilian Pantanal using camera-traps and capture-recapture sampling in combination with GPS radiotelemetry. Biol Conserv 29: 487- 496. doi: 10.1016/j.biocon.2005.11.023

46. Royle JA, Young KV (2008) A hierarchical model for spatial capture-recapture data. Ecology 89: 2281–2289. doi:10.1890/07-0601.1. PubMed: 18724738.

47. Royle JA, Karanth KU, Gopalaswamy AM, Kumar NS (2009) Bayesian inference in camera trapping studies for a class of spatial capture-recapture models. Ecology 90: 3233-3244. doi:10.1890/08-1481.1. PubMed: 19967878.

48. Gopalaswamy AM, Royle JA, Hines JE, Pallavi S, Jathanna D et al. (2011) SPACECAP: A Program to Estimate Animal Abundance and Density using Bayesian Spatially-Explicit Capture-Recapture Models. Wildlife Conservation Society - India Program, Centre for Wildlife Studies, Bengaluru, India. (Version 1.0.6).

49. R Development Core Team (2012) R: A Language and Environment for Statistical Computing. Vienna: R Foundation for Statistical Computing.

50. Royle JA, Magoun AJ, Gardner B, Valkenburg P, Lowell RE (2011) Density estimation in a wolverine population using spatial capture-recapture models. J Wildlife Manage 75: 604-611. doi:10.1002/jwmg.79.

51. Noss AJ, Gardner B, Maffei L, Cuéllar E, Montaño Ret al (2012) Comparison of density estimation methods for mammal populations with camera traps in the Kaa-Iya del Gran Chaco landscape. Anim Conserv 15: 527–535 doi:10.1111/j.1469-1795.2012.00545.x.

52. Tobler MW, Carrillo-Percastegui SE, Powell G (2009) Habitat use, activity patterns and use of mineral licks by five species of ungulate in South eastern Peru. J Trop Ecology 25: 261-270. doi:10.1017/S0266467409005896.

53. Carbone C, Christie S, Coulson T, Franklin N, Ginsberg J et al. (2001) The use of photographic rates to estimate densities of tigers and other cryptic mammals. Anim Conserv: 75-79. doi: 10.1017/s1367943001001081

54. Karanth KU (1995) Estimating tiger *Panthera tigris* populations from camera-trap data using capture-recapture models. Biol Conserv 71: 333-338. doi:10.1016/0006-3207(94)00057-W.
55. Chase Grey JN (2011) Leopard population dynamics, trophy hunting and conservation in the Soutpansberg Mountains, South Africa. PhD thesis, Durham University, UK.
56. Bailey TN (2005) The African Leopard: Ecology and behaviour of a solitary felid. New York: Columbia University Press.
57. Maputla NW, Chimimba CT, Ferreira SM (2013) Calibrating a camera trap-based biased mark-recapture sampling design to survey the leopard population in the N'wanetsi concession, Kruger National Park, South Africa. Afr J Ecol. doi: 10.1111/aje.12047
58. Herschel P (2008) The conservation biology of the leopard (*Panthera pardus*) in Gabon: Status, threats and strategies for conservation. PhD thesis. , University of. Göttingen, Germany.
59. Owen C (2006) Reproductive biology and population ecology of leopards (*Pantherapardus*) on Karongwe. MSc thesis. , University of KwaZulu-Natal, Durban University of. KwaZulu-Natal.
60. Balme GA, Batchelor A, de Woronin Britz N, Seymour G, Grover M et al. (2013) Reproductive success of female leopards *Panthera pardus*: the importance of top-down processes. Mammal Rev 43: 221–237. doi:10.1111/j.1365-2907.2012.00219.x.
61. Gerdel K (2008) Leopard (*Panthera pardus*) prey prediction using line transect sampling in the Western Soutpansberg mountain range, Limpopo province, South Africa. MSc. Thesis. , University of Tübingen, Germany.
62. Riley SJ, Malecki RA (2001) A landscape analysis of cougar distribution and abundance in Montana, USA. Environ Manage 28: 317–323. doi:10.1007/s0026702503. PubMed: 11531235.
63. Hatten JR, Averill-Murray A, Van Pelt WE (2005) A spatial model of potential jaguar habitat in Arizona. J Wildl Manage 69: 1024–1033. Available online at:doi:10.2193/0022-541X(2005)069[1024:ASMOPJ]2.0.CO; 2.
64. Gavashelishvili A, Lukarevskiy V (2008) Modelling the habitat requirements of leopard *Panthera pardus* in west and central. Asia - J Appl Eco 45: 579–588. doi: 10.1111/j.1365-2664.2007.01432.x

65. Willems EP, Barton RA, Hill RA (2009) Remotely sensed productivity, home range selection and local range use by an omnivorous primate. Behavioral Ecology 20: 985-992. doi:10.1093/beheco/arp087.
66. Willems EP, Hill RA (2009) Predator-specific landscapes of fear and resource distribution: effects on spatial range use. Ecology 90: 546-555. doi:10.1890/08-0765.1. PubMed: 19323238.
67. Lindsey PA, du Toit JT, Mills MGL (2005) Attitudes of ranchers towards African wild dogs *Lycaon pictus*: Conservation implications on private land. Biol Conserv 125: 113-121. doi:10.1016/j.biocon.2005.03.015.
68. Weaver JL, Paquet PC, Ruggiero LF (1996) Resilience and conservation of large carnivores in the Rocky Mountains. Conserv Biol 10: 964-976. doi:10.1046/j.1523-1739.1996.10040964.x.
69. Novaro AJ, Funes MC, Walker RS (2005) An empirical test of source–sink dynamics induced by hunting. J Appl Ecol 42: 910–990. doi:10.1111/j.1365-2664.2005.01067.x.
70. Robinson HS, Wielgus RB, Cooley HS, Cooley SW (2008) Sink populations in carnivore management: cougar demography and immigration in a hunted population. Ecol Appl 18: 1028–1037. doi:10.1890/07-0352.1. PubMed: 18536260.
71. Foster RJ, Harmsen BJ (2012) A critique of density estimation from camera-trap data. J Wildlife Manage 76: 224-236. doi:10.1002/jwmg.275.
72. Tobler MW, Carrillo-Percastegui SE, Zuniga Hartley A, Powell G (2013) High jaguar densities and large population sizes in the core habitat of the southwestern Amazon. Biol Conserv 159: 375-381. doi:10.1016/j.biocon.2012.12.012.
73. Tobler MW, Powell G (2013) Estimating jaguar densities with camera traps: Problems with current designs and recommendations for future studies. Biol Conserv 159: 109-118. doi:10.1016/j.biocon.2012.12.009.
74. Sollmann R, Furtado MM, Gardner B, Hofer H, Jácomo ATA et al. (2011) Improving density estimates for elusive carnivores: accounting for sex-specific detection and movements using spatial capture–recapture models for jaguars in central Brazil. Biol Conserv 144: 1017–1024. doi:10.1016/j.biocon.2010.12.011.

75. Ogada MO, Woodroffe R, Oguge NO, Frank LG (2003) Limiting depredation by African carnivores: the role of livestock husbandry. Conserv Biol 17: 1521-1530. doi:10.1111/j.1523-1739.2003.00061.x.
76. Jenny D (1996) Spatial organization of leopards (*Panthera pardus*) in Tai National Park, Ivory Coast: is rainforest habitat a "tropical haven"? J Zool Lond 240: 427- 440. doi:10.1111/j.1469-7998.1996.tb05296.x.
77. Hamilton PH (1976) The movements of leopards in Tsavo National Park, Kenya, as determined by radio-tracking. M.Sc. Thesis, University of Nairobi, Kenya.
78. Mizutani F (1999) Impact of leopards on a working ranch in Laikipia, Kenya. Afr J Ecol 37: 211-255. doi: 10.1046/j.1365-2028.1999.00171.x
79. Cavallo J (1993) A study of leopard behaviour and ecology in the Seronera valley, Serengeti National park. SWRC Serengeti Wildlife Research Centre Scientific Report. pp. 33-43.
80. Schaller GB (1972) The Serengeti Lion. Chicago: University of Chicago Press.
81. Stein AB, Fuller TK, DeStefano S, Marker LL (2011) Leopard population and home range estimates in north-central Namibia. Afr J Ecol 49: 383–387. doi:10.1111/j.1365-2028.2011.01267.x.
82. Stander PE, Haden PJ, Kaqece Ghau (1997) The ecology of asociality in Namibian leopards. J Zool Lond 242: 343-364. doi: 10.1111/j.1469-7998.1997.tb05806.x
83. Bothma J, du P, Le Richie E (1984) Aspects of the ecology and the behaviour of the leopard, *Panthera pardus*, in the Kalahari Desert. Koedoe (Suppl): 259-279. doi: 10.4102/koedoe.v27i2.585

Chapter 8

Climate Change and Challenges of Water and Food Security

Anil Kumar Misra*

Department of Civil and Environmental Engineering, ITM University, Sector 23A, Palam Vihar, Gurgaon 122017, Haryana, India

ABSTRACT

Water and food security are the key challenges under climate change as both are highly vulnerable to continuously changing climatic patterns. Studies have predicted that the average global temperature may increase by 1.4–5.8°C and there would be substantial reduction in fresh water resources and agricultural yield by the end of the 21st century. Approximately 75% of the Himalayan glaciers are on retreat and will disappear by 2035. Moreover in Africa (Sub-Saharan Africa) by 2050 the rainfall could drop by 10%, which would reduce drainage by 17%. Majority of the fresh water resources has already been depleted and there is reduction in agricultural production globally with escalation in population and food demand. Some of the prominent climate change impacts are, growing deserts, and increase in the magnitude of floods and droughts. An extreme decline in crop yields in arid and semi arid areas globally has caused food shortages and a manifold increase in food inflation. Countries of Africa, Middle East, Arab and Asia have close economic ties with natural resource and climate-dependent sectors such as forestry, agriculture, water, and fisheries. This manuscript highlights groundwater recharge by utilization of wastewater using the Soil Aquifer Treatment (SAT) method in irrigation and the significance and methods of artificial recharge of groundwater. This paper also presents easily and economically feasible options to ensure water and food security under climate change and recommend formation of effective adaptation and mitigation polices and strategies to minimizing the impact of climate change on water resources and irrigation.

INTRODUCTION

Water and food scarcity are the biggest problem globally and it severely affects the arid and semiarid regions/countries. Climate change has resulted in increases in globally-averaged mean annual air temperature and variations in regional precipitation and these changes are expected to continue and intensify in the future (Solomon et al., 2007). The impact of climate change on the quantity and quality of groundwater resources is of global importance because 1.5–3 billion people rely on groundwater as a drinking water source (Kundzewicz and Döll, 2009). As per the fourth IPCC assessment report the knowledge of groundwater recharge and of levels in both developed and developing countries is poor. There has been very little research on the impact of climate change on groundwater' (Kundzewicz et al., 2007).

Study of Global Climate Models (GCMs) projects significant changes to regional and globally averaged precipitation and air temperature, and these changes will likely have associated impacts on groundwater recharge (Kurylyk and MacQuarrie, 2013). IPCC report

(2008) predicts that the climate change over the next century will affect rainfall pattern, river flows and sea levels all over the world. Studies show that agriculture yield will likely be severely affected over the next hundred years due to unprecedented rates of changes in the climate system (Jarvis et al., 2010 and Thornton et al., 2011). In arid and semi-arid areas the expected precipitation decreases over the next century would be 20% or more. The accelerated increase in the greenhouse gases (GHG) concentration in the atmosphere is a major cause for climate change. As per theIPCC (2007) report, the maximum growth in the emission of greenhouse gases (GHG) has occurred between 1970 and 2004, i.e. 145% increase from energy supply sector, 120% from transport, 65% from industry, 40% from change in land use patterns and during this period global population increases by 69%. As per the WMO (2013), the world experienced unprecedented high-impact climate extremes during the 2001–2010 decade that was the warmest since the start of modern measurements in 1850. Moreover, survey of 139 National Meteorological and Hydrological Services and socio-economic data and analysis from several UN agencies and partners conducted by WMO concluded that floods were the most frequently experienced extreme events over the course of the decade. The amount of energy reaching the earth's atmosphere every second on a surface area of one square meter facing the sun during daytime is about 1370 Watts and the amount of energy per square meter per second averaged over the entire planet is one quarter of this (IPCC, 2007A). The global mean temperature has increased by 0.74 °C during (Fig. 1) the last 100 years. Furthermore studies conducted by Indian Space Research Organization (ISRO) after the study of 2190 Himalayan glaciers revealed that approximately 75% of the Himalayan glaciers are on the retreat, with the average shrinkage of 3.75 km during the last 15 years (Misra, 2013). These findings raise serious concerns over the accelerated retreat of glaciers in the Himalayan Mountains because it will increase the variability of water flows to downstream regions and threaten the sustainable water use planning in the world's most populous Ganga Basin. Studies (de Wit and Stankiewicz, 2006 and Anthony Nyong, 2005) predict that by the year 2050 the rainfall in Sub-Saharan Africa could drop by 10%, which will cause a major water shortage. This 10% decrease in precipitation would reduce drainage by 17% and the regions which are receiving 500–600 mm/year rainfall will experience a reduction by 50–30% respectively in the surface drainage.

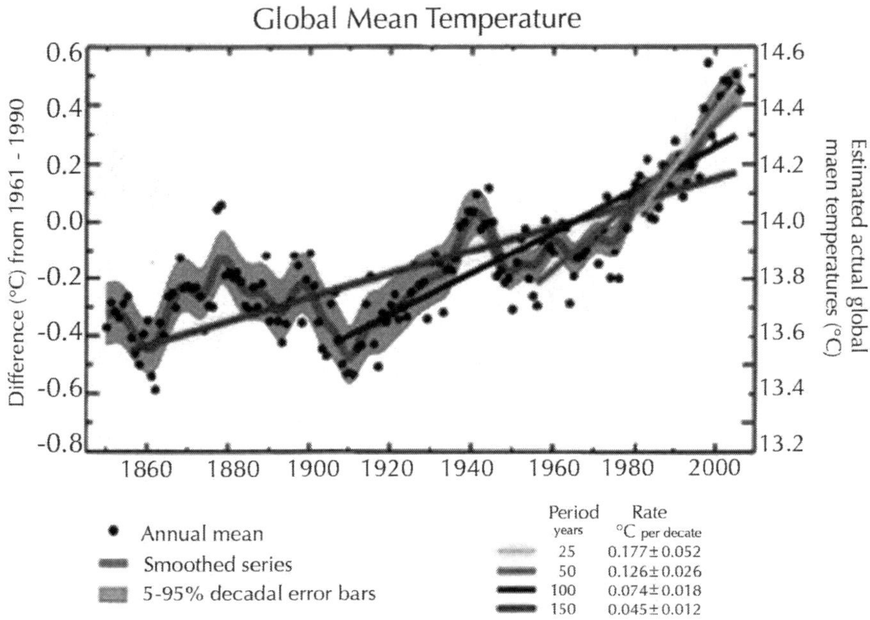

Figure 1: Global mean temperature during the last 100 years (source: IPCC, 2007A).

So, far much attention has been given to climate change adaptation as an anticipatory and planned process, managed through new policies, technological innovations and development interventions (Adger et al., 2005). But these policies and strategies are far from implementation and most of the fresh water resources are depleting at a very fast rate due to unprecedented escalation in demand from domestic, irrigational and industrial sectors. Impact of climate change such as depletion of water resources (Shallow & deep aquifer depletion) and decline in agricultural production has increased and has escalated food inflation globally and there is an acute shortage of food in many poor African and Asian countries, where people cannot afford expensive food and are dying of starvation. The condition is extremely severe in continents like Africa, where most of the northern portion is extremely dry. Western India, Middle East and Arab Countries, where most of the domestic, irrigational and industrial demands are met by Surface and groundwater are facing severe crisis due to depletion of water resources.

This condition can only be improved by increasing the crop yield and preventing further depletion of water resources. The paper highlights the best suitable methods, which are easily and economically feasible and can ensure water and food security under climate change if implemented properly. The manuscript also suggests a road map for long-term and near-term strategy for minimizing the impact of climate change on water resources and agriculture.

PRESENT STATUS OF WATER RESOURCES

There are large uncertainties among the nations vulnerable to climate change impacts over the availability of water resources in the future. In 1955, only seven countries were found to be with water stressed conditions. In 1990 this number rose to 20 and it is expected that by the year 2025 another 10–15 countries shall be added to this list. It is further predicted that by 2050, 2/3rds of the world population may face water stressed conditions (Gosain et al., 2006). Majority of the Arab countries depend on the international water bodies for their requirement. Arab countries do not have any major source of water; they have to depend on natural precipitation and water conservation techniques. Nile river basin is the home of approximately 190 million people of Ethiopia, Eritrea, Uganda, Rwanda, Burundi, Congo, Tanzania, Kanya, Sudan and Egypt. Since majority of nations of the Nile river basin are among the top 10 poorest countries of the world therefore it is absolutely difficult for them to adopt any strategy of water management, which require investment. The Middle East and the OSS (Observatory of the Sahara and the Sahel) regions, which have the least natural water resources, both in absolute terms and in relation to its population (UNESCO, 2004) will be affected severely. Table 1 shows the world's natural renewable water resources as compared to OSS region.

Table 1: World's natural renewable resources (rounded off figures) (source: UNESCO, 2004)

Geopolitical regions (Groups of Countries)	Total average internal and external resources (km^3/year)	Portion coming from outside of the region (external resources) (km^3/year)	Relatively constant portion (Surface and groundwater) (km^3/year)
OSS region**	520	113	~200
Europe	1900	10	600
Ex-USSR (the former USSR)	4400	430	1400
North America (USA and Canada)	6700	0	1700
Latin America (including the Caribbean)	13,000	3	4000
Africa, excluding the OSS region (but including the Madagascar)	3500	0	1200
Near the Middle East	480	17	100
Indian sub-continent and south east Asia	6600	1000	1600
China (including Mongolia and North Korea)	2800	0	1000
Japan and 'four dragons'	700	0	200
Australia and Oceania	2000	0	300
Total (without overlap)	42,600		12,300

OSS region**: Arabic Maghreb Union (AMU) countries: Algeria, Egypt, Libya, Mauritania, Morocco and Tunisia. Permanent Interstate committee for drought Control in the Sahel (CILSS) countries: Burkina faso, Cape Verde, Cad, Gambia, Guinea-Bissau, Mali, Mauritania, Niger and Senegal. Intergovernmental authority on Development (IGAD) countries: Djibouti, Eritrea, Ethiopia, Kenya, Somalia, Sudan and Uganda.

The use of strong technological equipment such as deep tube wells and high-powered pumps for the abstraction of groundwater in majority of the vulnerable countries resulted in continuous unsustainable drawdown of aquifers. These pumps allowed faster drafting from aquifers, rivers, canals etc. and disturb the natural equilibrium of recharge and discharge. The country like India where the population is increasing in an unprecedented rate is likely to be water scarce by 2050. The water requirement in India by 2050 will be in the order of 1450 km^3, which is significantly higher than the estimated water resources of 1122 km^3 per year. Therefore to meet the shortfall requirement, it is necessary to harness additional 950 km^3 per year over the present availability of 500 km^3 per year (Gupta and Deshpande, 2004). Table 2 shows the basin wise water present in India. As per the study of the Ministry of Water Resource (MOWR) Govt. of India, the estimated irrigation return flow (RF) from the surface and groundwater irrigation is likely to be 223 km^3 per year in the year 2050 for higher population growth rates giving 133 km^3 per year. The total recyclable wastewater is estimated to be 177 km^3 per year in 2050. Taking all these important factors into consideration, Gupta and Deshpande (2004) estimated the total resource availability in 2050 for higher population growth shown in Table 3.

Table 2: Basin wise water in India (km^3/year) source: MOWR (1999)

Sl. No.	Basin	ASW	AMR	EUSW	RGW	SGW
1	Indus	73.30	58.60	46.00	26.50	1338.20
2a	Ganga	525.00	401.30	250.00	171.60	7834.10
2 (b + c)	Brahmaputra + Meghana	585.70	477.50	24.00	35.10	1018.50
3	Godavari	111.40	107.10	76.30	40.60	59.40
4	Krishna	78.10	61.00	58.00	26.40	36.00
5	Cauvery	21.60	18.90	19.00	12.30	42.40
6	Pennar	6.70	6.20	6.70	4.90	11.10
7	EF : Between Mahanadi & Pennar	33.50	15.30	13.10	18.80	41.30
8	EF: Between Pennar & K. Kumari	16.50	16.00	16.50	18.20	66.00
9	Mahanadi	66.90	60.20	50.00	16.50	119.70
10	Brahmani – Baitarni	33.00	32.60	18.30	4.10	43.40
11	Subarnaretha	12.80	9.70	6.80	1.80	10.80
12	Sabarmati	3.80	3.40	1.90	3.20	28.20

13	Mahi	11.00	10.70	3.10	4.00	12.60
14	WF: Kutchh, S'tra, luni	15.10	13.60	15.00	11.20	113.20
15	Narmada	46.00	36.90	27.50	10.80	18.40
16	Tapi	16.90	16.20	15.00	8.30	7.50
17	WF: Tapi & Tadri	87.40	80.30	11.90	17.70	11.20
18	WF: Tadri to K. Kumari	113.50	97.80	24.30	0.00	0.00
19	Inland drainage: Rajasthan	0.00	0.00	0.00	0.00	0.00
20	MR: B'desh and Myanmar	31.00	24.80	0.00	0.00	0.00
Total		1889.20	1548.10	683.40	432.00	10812.00

EF: East Flowing; WF: West Flowing; MR: Minor Rivers; K. Kumari: Kanyakumari; S'tra: Saurashtra; B'desh: Bangladesh; Available Surface Water (ASW); Average Monsoon Runoff (AMR); Economically Utilizable Surface Water (EUSW); Replenishable Groundwater (RGW); Static Reserve of Groundwater (SGW).

Table 3: Water resources availability based on low and high population growths for 2050 (km^3); source: Gupta and Deshpande (2004)

Water available during 2001	Water required during 2050	Anticipated water deficit	Possible measures to meet the deficit				
			EUSW+GW in excess of 1998	Recyclable waste water	Irrigation return flow	RAGWR	Water availability
500	973–1450	473–950	SW = 420	103–177	33–133	125	1311–1485
			GW = 202				
			Total = 550				

RAGWR = Retrievable Artificial Groundwater Recharge; EUSW = Economically Utilizable Surface Water; GW = Groundwater.

- Ignored water quality issues.
- After considering 17% decline in storage for surface sedimentation.

Africa which is one of the world's driest continents is facing a very severe water crisis. Over 90% of Sub-Saharan Africa agriculture is rain-fed, and mainly under smallholder management (Batino and Waswa, 2011). As per the WHO/UNICEF report more than one billion people still use unsafe drinking water. The African river system identifies three river systems (de Wit and Stankiewicz, 2006) i.e. the areas receiving very low rainfall have virtually no perennial drainage (dry regime), then the areas with an intermediate range in which drainage density increases with increasing rainfall (intermediate rainfall regime) and the areas of high rainfall (high rainfall regime). The dry regime covers the largest area of the African continent i.e. approximately 41%, but most important is the intermediate rainfall regime which covers approximately 25% because this is the area where changes to precipitation would result in serious changes in drainage supply. Further as predicted by an ensemble of global climate change models by the second part of this century, climate change would directly affect African countries, 75% of which belongs to the intermediate stage. Fig. 2 shows the present rainfall regimes in t Africa and Fig. 3 shows the expected changes in the precipitation by the end of the 21st century on the basis of the composite of 21 leading fully coupled GCMs adapted by IPCC for forecasting purposes (CSAG, 2002). The studies conducted by several researchers predict that the crop yield will decline and the crop water demand will increase in the African continent especially in the dry land farms. A net 2.5 °C rise in temperature in Africa will result in a decline of net revenues from agriculture by US$ 23 billion (Kurukulasuriya and Mendelson, 2007). Thus it has become necessary now to take very seriously the impact of climate change on the present water resources and take necessary actions without any further delay.

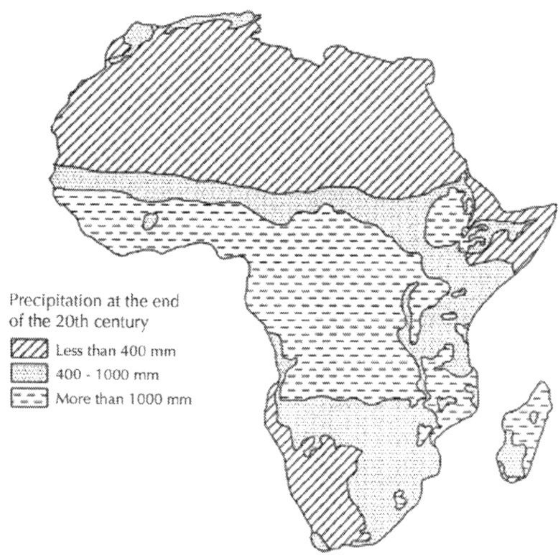

Figure 2: Precipitation in the African continent at the end of 20th century (source: de Wit and Stankiewicz, 2006).

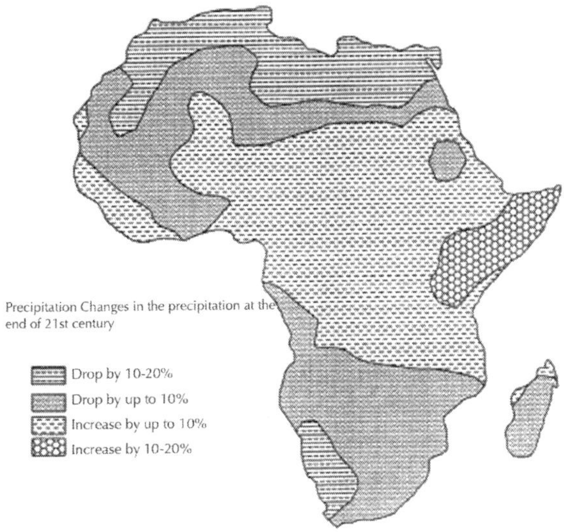

Figure 3: Predicted changes in the precipitation in the African continent due to climate change at the end of the 21st century (source: de Wit and Stankiewicz, 2006).

STRATEGIC CHALLENGES

Water conservation is one of the oldest activities practiced by old civilization to fulfill the required demand for water for irrigated agriculture and domestic needs in the arid and semi-arid regions. In ancient times the recharge movement initiated by the local communities was aided and supported by emperors. The approaches that support farming communities to self-mobilize and self-organize for participatory learning and action could lead farmers to enhance their uptake of better technologies and improved use of farm-level resources in the wake of increased climate change and variability (Mapfumo et al., 2013). One of the most important facts of water resource management is that the conservation of water resources costs far less and it is more sustainable than treating non-potable and waste water and supplying it for the required needs. Further majority of water conservations strategies are relatively easily and economically feasible as compared to the water treatment plants. Therefore good strategies for artificial recharge to groundwater and conservation of water resources are necessary. Further, the irrigation system must be designed, installed, managed, and maintained properly, because it wastes lot of water. Modern irrigational equipment should be used and their regular maintenance is essential. Table 4 shows the performance of some of the water management strategies after considering global warming effects.

Table 4: Performance of water management strategies after considering the Global warming effects (source: Nelson, 2007)

More effective	Not affected	Less effective
Landscape conservation	Wastewater recycling	Traditional river diversions
Conservation rate structures	Interior water conservation	Traditional groundwater pumping
Agricultural water conservation	Groundwater cleanup	Traditional surface storage facilities
Water marketing		Ocean water desalination
Urban storm water management		
Saltwater groundwater intrusion barriers to protect coastal aquifers		
Water system reoperation		
Interagency collaboration and integrated water management strategies		
Floodplain management		
Watershed Restoration		

Given existing energy requirements.

There are several key challenges related with policy and strategy making that have to be confronted. Some of the important challenges are as follows:-

(i) Collection of information and data and their sharing among countries related with climate change and its impact on water resources. It is necessary because water resource management requires, systematic and well planned actions based on accurate scientific data.

(ii) Majority of the countries have varying hydrological conditions, therefore adaptation of the same policies and strategies by each country is not possible. The policies and the action plan will be different for each country based on its hydrological conditions.

(iii) Climate change has increased the frequency and intensity of the natural calamities and now it has become necessary to invest in the study of these natural calamities and their future impacts and prepare a comprehensive plan to minimize their impact on countries.

(iv) Among all the challenges the biggest challenge is the financing of the climate change study related projects and the adaptation of their recommendation because these investments are not profitable.

(v) Many poor and developing countries are unable/hesitant to construct expensive infrastructure required to meet climate change related challenges because of imperfect information and data about intensity of climate change impact on their country.

To minimize the impact of climate change on water resources it is necessary to understand and evaluate the vulnerability of water resources to global warming impacts. After understanding these impacts only policies and strategies should be formed and implemented. All efforts should be made to present future impacts by reducing greenhouse gas emission

WATER SECURITY UNDER CLIMATE CHANGE

In most developing countries, especially African and Asian, there are urgent needs to understand the dynamics of local climate and make predictions to respond to climate variability and change. The economies of most developing countries depend heavily on climate-sensitive sectors such as water, agriculture, fisheries, energy and tourism, climate change therefore poses a serious challenge to social and economic development in developing countries. (Munang et al., 2013). The shortage of water can be augmented from wastewater utilization after suitable treatment (FAO, 2012). Some of the techniques like artificial recharge and use of unconventional water (reuse of wastewater after recycling) are effective solutions to minimize the impact of such problems. The unconventional water can play a major role in the management of water resources and agricultural activities in dried and semi dried areas. Recycling and reuse of wastewater for beneficial purposes such as agricultural and landscape irrigation, industrial processes and replenishing a groundwater aquifer (groundwater recharge) can help in minimizing the impact of climate change on crop yield and water resources. Recycled water for irrigation requires less treatment than recycled water for domestic purposes and till date no documented case of human health problems has been reported by the use of unconventional water for irrigational purposes. Through the natural water cycle the earth has recycled and reused water for millions of years. Water recycling by giving technological support can speed up these natural processes. Usually the recycling of unconventional water may be classified as planned recycling and unplanned recycling.

Unplanned Recycling

It occurs usually in the cities located near the banks of the rivers. These cities take water from rivers to fulfill their demands of domestic and industrial purposes. Cities discharge their wastewater in downstream and draw water from upstream and after recycling it is reused.

Planned Recycling

Planned recycling of wastewater is developed with the goal of beneficially reusing a recycled water supply. Many cities established a water recycling master plan to utilize precious water resources. Under such a plan, all the water-related measures are examined for wastewater recycling and all activities within the city are carried out (establishment of efficient water and wastewater systems within the city) following the master plan. The objectives of such recycling master plan include the creation and nurturing of a water cycle that has a minimal impact on the environment.

Another important way of reusing the unconventional water is through artificial recharge of the dried aquifer systems with partially treated water. Such type of aquifer recharge is only possible where the subsurface lithological conditions and the groundwater condition are favorable, a high degree of upgrading can be achieved by allowing sewage effluent to infiltrate into the soil and move down to the groundwater. The unsaturated lithology (Vadose Zone) then acts as a natural filter and can remove essentially all suspended solids, biodegradable materials, bacteria, viruses and other microorganisms (Fadlelmawla et al., 1999).

This type of artificial recharge to groundwater with wastewater can be very useful to prevent depletion of groundwater in arid and semiarid areas, where precipitation is extremely low and almost no rechargeable water is available. This process is also called Soil-Aquifer Treatment (SAT). Several practical applications of SAT technology are available throughout the world. One of the biggest is at Whittier Narrows, Los Angeles, USA, which has a recharge area of approximately 279 ha forms a small basin is divided into 15 sub-basins. The scheme disposes of 750 Ml/day using only about one third of the basin during normal operation. The major objective of the scheme is to replenish groundwater for local abstraction and only 16% of the recharge is renovated wastewater. No measurable impact has been detected on either groundwater quality or human health (Nellor et al., 1985).

Artificial recharge with wastewater is totally depending on the properties of the subsurface lithological conditions. The areas where Vadose Zones are not supportive (hard rock areas, older alluvial plain with thick clay and silt layers) for groundwater recharge, artificial recharge structures, like horizontal shaft and vertical shaft, equipped with artificial set of lithologies (Fig. 4 and Fig. 5) can be very effective. Artificial lithologies are equipped with geotextiles that perform both separation and filtration functions and improve the performance of recharge structures manifold. These structures usually vary in shape and size and can be used in construction in hard rock and older alluvial plains also. These structures are very effective if formed up to the depth of shallow aquifers (Misra et al., 2013). Moreover the purification mechanisms through an artificial set of lithology is much better than the natural conditions.

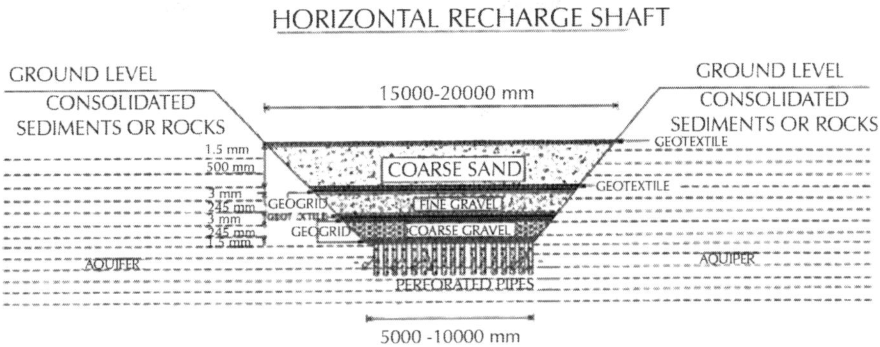

Figure 4: Design of horizontal shaft structure equipped with artificial set of lithology suitable for hard rock areas, older alluvial plains and places with high surface runoff.

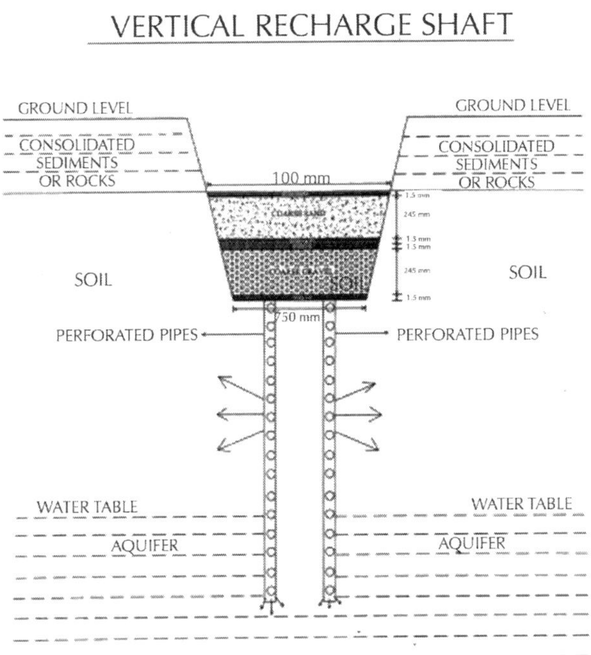

Figure 5: Design of vertical shaft structure equipped with artificial set of lithology suitable for hard rock areas and older alluvial plains and places with low surface runoff.

Utilization of Wastewater for Groundwater Recharge

The artificial recharge to groundwater with wastewater depends on a number of factors such as porosity, permeability, hydraulic conductivity, transmissivity and the most important stratigraphical arrangement. In majority of the cases the subsurface sequencing of the layers and the aquifer depth creates problems. Thus the selection of the possible locations for the construction of artificial recharge facility should be established taking care of factors such as suitable site, infiltration capacity, hydrogeological conditions, systematic planning and environmental considerations. Based on these considerations Soil-Aquifer Treatment (SAT) facility can be established.

Identification of Suitable Area

Identification of a suitable area for such recharge schemes should be demarcated very carefully. Such areas should be as far as possible with any urban settlement and be a micro or mini watershed. Demarcation of areas should be based on the following criteria. (a) A minimum of 20 m depth to the groundwater levels with continuous declining trends is required to allow for geopurification processes (infiltration and absorption) before the infiltrating wastewater reaches the groundwater. This much of depth also allows for groundwater mounding during the recharge process without affecting the infiltration process (RIGW, 1999). (b) Where groundwater quality is good and there is no alternative source of water (c) where discharge from wells and hand pumps are inadequate.

Infiltration Capacity of the Vadose Zone

Infiltration capacity of the Vadose Zone is a very important factor that governs the rate of saturation of the zone of aeration. Therefore the lithological zones with an infiltration rate of 0.30 m/day or more are most suitable.

Hydrogeological Consideration

Integration of data obtained from field, laboratory analysis and simulation methods are generally used to develop an understanding of the hydrogeological system as a basis for predicting potential consequences. Some of the important considerations that are important for such projects are as follows

(i) The aquifer characteristics at the recharging site must have good hydrogeological conditions such as storage coefficient, availability of storage space and permeability.

(ii) Young alluvium buried channels, alluvial fans and sand dunes etc. are some of the best places for recharge.

(iii) In hilly terrain and hard rock areas, fractured, weathered to semi weathered and cavernous rocks are the best sites for recharge.

Planning & Management Consideration

Planning and management of artificial recharge projects need careful consideration of several water management objectives, water routing capabilities, economics, off-site effects and different other factors. Water requirement for irrigation purposes can be fulfilled by geopurified wastewater, if it is carried out with systematic planning. A large amount of wastewater is generated in urban areas. After partial treatment this water can be used for artificial recharge of groundwater aquifer, which can be exploited for irrigational purposes. For creating partial treatment facility near the recharge and discharge area planning should be based on hydrogeological consideration and based on easily and economically feasible techniques.

Appropriate Spacing

It is also important that there should be appropriate spacing between the two recharge structures. Because when the recharge water begins to merge with the existing groundwater, a mound will develop that drastically reduces the recharge rate. Further if many recharge structures are placed side by side the hydraulic gradients of these recharges will become zero due to superposition, due to this the mound will continue to grow and the recharge rate will drop more quickly (Chatdarang,

2001). Therefore two recharge structures should be at least 1000 m away from each other for getting good results.

Environmental Consideration

It is very important to carry out a detailed environmental impact assessment of each selected location and project before the implementation, which includes mitigation and monitoring planning. The most important factor for environmental consideration is that in any circumstances the selected site should not be within or upstream of a groundwater drinking community and no recharge should be planned where groundwater is flowing into the river.

Artificial Recharge Methods

Artificial recharge methods were specially design for uses in the arid and semi arid regions of the world. These recharge methods include three (Sakthivadivel, 2007) important components (i) Area of the watershed to produce runoff (ii) a storage facility (soil profile, surface reservoirs or groundwater aquifers); and (iii) a target area to beneficially use the water (agriculture, domestic or industry). According to O'Hare et al. (1986)some of the important factors, which are considered for artificial recharge are (a) Availability of wastewater (b) Quantity of source water available (c) Resultant water quality (after reactions with native water and aquifer materials) (d) clogging potentials (e) Underground storage space available (f) Depth of underground storage space (g) Transmission characteristics of the aquifer (h) Applicable methods (injection or infiltration) (i) Legal/institutional constraints (j) Costs (h) Cultural/social considerations. The programs of the artificial recharge are usually conducted in four phases.

Feasibility Studies

The feasibility study is the detailed evaluation of the entire recharge project that includes planning, demarcation of the area for recharge, groundwater levels and their declining trend, precipitation over the demarcated area, available quantity of water, study of the geological conditions, hydrological variability within the aquifer, chemical mixing of surface waters with native groundwater, the nature of probable migration of recharged water and economical feasibility of the entire project.

Analytical and Testing

Analytical and testing programs are based on the results of the feasibility analysis. The test programs are usually designed, using different existing facilities. This work includes, analysis and testing of quality of water and the pretreatment required, conveyance system required to bring the water to the recharge site, chemical and physical modeling of recharge options, detailed chemical analysis of co-mingled waters that have different initial chemical signatures, measurement of infiltration capacity of soil and recharge rates.

Designing and Operation

Design of the recharge structure is based on the outcome of the feasibilities, analytical and testing studies. If the design is not compatible with the existing ground environment, then the operation of the recharge structure will completely stop within few days. This phase includes the works like designing of injection wells, induced recharge structures; recharge shafts with filter, and different land improvement and watershed management techniques.

Project Implementation

Successful implementation of any recharge project is based on the results of the above mentioned three phases of the project. Usually project implementation phase consists of full scale program parameters that include site for well/shaft/infiltration ponds if required, major future options for sourcing of surface water, planning of different recharge management techniques during regular operations and continuous monitoring and maintenance of entire working mechanism of the project.

Cost of Recharge Structures

Cost of any recharge structure depends upon the source water treatment, source water transportation, resistance to siltation and clogging and the most important stability of recharge structure. The cost of construction, operation, monitoring and maintenance usually depends on the design of the recharge structure and its compatibility with the existing ground environment. Generally the construction and operational costs of the recharge structure is far less than those required for domestic water supply using tankers. Table 5shows the capital cost and operational cost of various recharge structures.

Table 5: Construction and operational costs of Artificial recharge methods (source: UNEP International Environment Centre, 2004)

Artificial recharge structure type	Capital cost/1000 m^3 of recharge structure	Operational cost/1000 m^3/year
Injection well (alluvial area)	$551	$21
Injection well (hard rock)	$2	$5
Spreading channel (alluvial area)	$8	$20
Recharge pit (alluvial area)	$515	$2
Recharge pond or percolation pond (alluvial area)	$1	$1
Percolation tank (hard rock area)	$5	$1
Check dam	$1	$1

Environmental Impact

Usually recharge structures do not show any harmful environmental affects if its design, operation, maintenance and monitoring are carried out following standard feasibility study recommendations. While planning the recharge structures, the environmental considerations include, ecological effects on soil, hydrologic and aquatic ecosystems, effects on the species and most important the possible effects on people's use of water resources for recreation. Majority of the recharge methods which use surface infiltration systems for recharge and depend on subsurface lithological infiltration rate are very susceptible to clogging by suspended particles, chemical impurities and biological activities. The clogged top layer of the rechargeable ground surface reduces its infiltration rate. Thus clogging of infiltration systems of the artificial recharge structures is a major operational problem. The clogging causes water logging problems that led to several environmental hazards such as soil salinity, loss of vegetation covers/irrigated area. Further due to water logging, water borne diseases, infectious diseases and several diseases spread by mosquitoes are common and often create panic situations.

Rain water Harvesting

In rain water harvesting the source of recharge water is rainfall only, while in the case of artificial recharge structures it could be rainwater, surface runoff, river or canal water or wastewater. Rain water harvesting can be defined as receiving, collection, storage, and use of rainwater for domestic, industrial and irrigational purposes. Besides its application domestic, industrial and irrigational purposes it may be effectively utilized to recharge groundwater aquifers.

Several types of rainwater harvesting structures can be constructed based on the type of catchment surface used (after Kahinda and Taigbenu): (1) in situ rainwater harvesting structures, where the system uses part of the target area as the catchment area, for instance construction of small ridge and furrow structures in agriculture fields (2) ex situ, rainwater harvesting structures where the system uses an uncultivated area as its catchment area, for instance diversion of surface runoff water into a collection basin (3) domestic rainwater harvesting structures, where the system collects water from rooftops, courtyards, compacted or treated surfaces, store it in tanks for domestic uses.

AGRICULTURE AND FOOD SECURITY

According to the Food and Agriculture Organization (FAO) of the United Nations, in most recent estimates, the number of people suffering from chronic hunger has increased from under 800 million in 1996 to over a billion (FAO, 2009). Majority of the world's hungry population are in South Asia and Sub-Saharan Africa. These regions have large rural populations, widespread poverty and extensive areas of low agricultural productivity due to steadily degrading resource bases, weak markets and high climatic risks (Vermeulen et al., 2012).

As per the current knowledge and options for supporting farmers, particularly smallholder farmers, in achieving food security through agriculture under climate change actions in following (Vermeulen et al., 2012) areas should be taken:

(a) Accelerated adaptation to progressive climate change over decadal time scales.

(b) Management of agricultural risks associated with increasing climate variability and extreme events.

(c) Mitigation actions that involve both carbon sequestration and reduction of emissions.

Adaptation of Advance Agriculture Practices

Adaptation of advance technologies can help in making the framing more easy and productive task. The good agricultural practice involves providing adequate water to plants so that excess standing water can be prevented. Excessive amounts of water can cause poor aeration of the root system leading to inhibition of plant development; therefore avoiding water stress is particularly important in arid regions. Further use of GPS tractors, along with sprayers can minimize the farming cost. These tractors can accurately drive themselves through the field after receiving the correct work plan in their computer system. Moreover the pressurized irrigation with sprinklers is extremely useful in increasing the water use efficiency in modernizing the agriculture system. Furthermore adaptation of mobile technique can play a big role in monitoring and controlling crop irrigation systems. With these instruments a farmer can control his irrigation systems from a phone or computer instead of driving to each field. The major challenge is to accelerate the adaptation of modern techniques in irrigation with which the crop productivity cannot be increased. Fig. 6 shows the framework for agricultural water management in the context of the climate change.

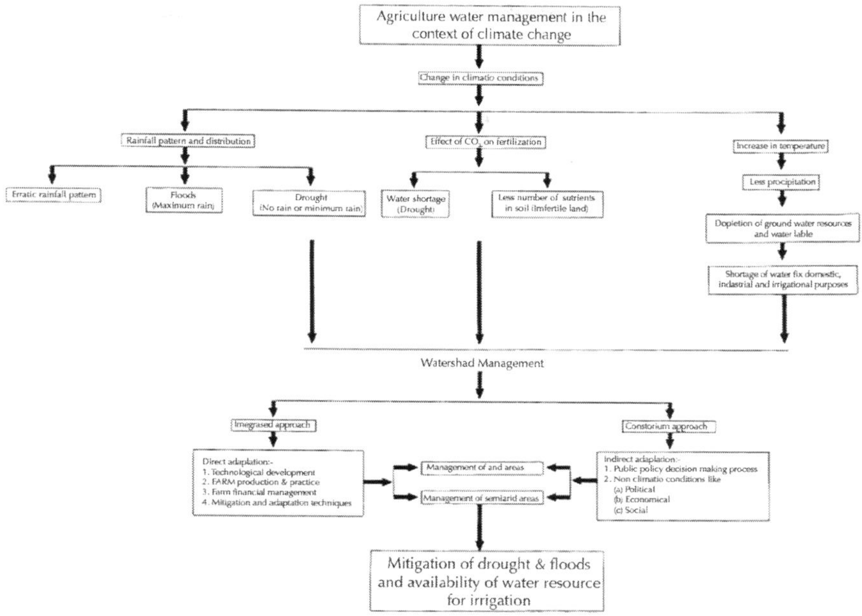

Figure 6: Proposed framework for agricultural water management in the context of climate change.

Crop Breeding

Adaptation of plant breeding can improve the quality, performance and diversity of crops with the objective of developing plants better adaptable to human needs. Studies show that for overcoming abiotic stresses in crops, the crop breeding technique has been proved to be an effective means of increasing food production (Evenson and Gollin, 2003), and mitigating climate change effects (Burney et al., 2010). Crop breeding techniques not only enhance the crop quality and yield and but also increase the tolerance of environmental pressures like soil salinity, high temperature and drought conditions. Therefore large scale adaptation of crop breeding is essential to ensure food security.

Climate Forecasting for Crops

Erratic climatic conditions and their variability with time play an important role in the crop production and overall yield. Most of the crop failures worldwide are associated with either a lack or excess of rainfall. Precise climate forecasting can reduce the risks of crop failure and also help in the pre and post decision making processes for better agricultural yield. Several studies have evaluated the potential benefits of using seasonal climate forecasts on the decision making process in agriculture (Jones et al., 2003 and Hansen, 2002). Further the nature of the forecasting also influences the ability of farmers to respond like farmers are more concerned about within-season characteristics of rainfall rather than the total seasonal rainfall amounts. The value of forecasts diminishes if information is received after the number of pre-planting decisions are made, therefore the forecasting should be in time and specific.

Managing Food Security

Managing food security and its sustainable development is one of the biggest challenges worldwide. Majority of the world's poor population lives in South Asia and Sub-Saharan Africa countries, nearly 1.7 billion people (Chen and Ravallion, 2007) and out of this, approximately for 860 million people there food security. Countries are not able to provide sufficient quantities of nutritious food to these people so that they can live healthily. It is strongly believed that there is enough food in the world to feed everyone adequately but the problem is distribution and management. Therefore development and implementation of a food security plan is necessary that must include procedures for handling threats, product tampering, and product storage and distribution plan along with a monitoring procedure. Moreover there should be a corrective action that prevents products from entering commerce.

Reduction in Farming Cost

Artificial recharge can play a major role in minimizing the farming cost. There are a number of examples that proof its utility in irrigation sector. In 2000, the International Water Management Institute (IWMI) carried out a study in India on the project "Madhya Ganga Canal Project (MGCP)", which occupies lower Ganga canal command area in central ganga plain. The study was carried out on Lakhaoti Branch canal of the MGCP, which spreads over 205.6 thousand hectares and covers the area of western Uttar Pradesh (UP) in India. Under this experiment excess river water was used to recharge the groundwater via earthen canals. The experiment was successful in raising the groundwater table to 6.6 meters and brought down the cost of pumping for irrigation from Rs. 4500 per ha meter to an economical level of Rs. 2700 per ha meter. The conclusion of this study is that the prevailing practices of supplying water only during summer/dry seasons need to be changed. In fact during monsoon such supplies should be carried out to help farmers to grow water intensive crops, this way both irrigation and recharge of groundwater will take place.

The construction of structures like "ooranies" is another example. These are big or small types of dugout tanks and ponds constructed thousands of years ago in very large numbers (more than 500,000) in the Peninsular India. These "ooranies" were constructed to fulfill the demand for water for agriculture and domestic purposes. There are several shallow dug wells near these tanks which are recharged with tank water. Further many wells which are used for drinking purposes, located near the tank and tank beds are artificially recharged from the tank into these wells to provide a clean water supply throughout the year with natural filtering (DHAN foundation, 2002). The structures like "ooranies" if formed in large scales especially in the arid and semiarid areas, the problems of water scarcity for domestic and irrigational purposes can be solved up to some extent despite drastic changes in climate.

CONCLUSION AND RECOMMENDATION

Climate changes have started showing its impact on water resources and agricultural yield worldwide. Majority of the countries in arid and semiarid areas totally depend on precipitation and rivers originating in tropical and temperate regions. The overall water stress is continuously increasing and due to climate change a sharp decline in precipitation is expected in these regions. Studies also predict reduction in frequency and escalation in the intensity of rainfall, which will result in frequent drought and floods. Unsustainable depletion of groundwater will likely be worsened by reduced surface water infiltration in arid and semiarid areas and the increase in intrusion of salt water to coastal aquifers from sea level rise will further reduce the availability of usable groundwater.

Agricultural sector and food securities are threatened and if the basic adaptive measures such as changes in crop pattern, crop breeding and types and innovative technologies, which use less water are not used global food production especially in arid and semi-arid areas will further decline. The present situation in the majority of the arid and semiarid countries is not satisfactory. These countries are not able to fulfill the required demand for water and food for people. The implementation of recycling and reuse of wastewater is a good option in these countries. Groundwater recharge using artificial recharge structures and Soil Aquifer Treatment (SAT) systems equipped artificial set of lithologies through wastewater can help in minimizing the impact of climate change on water resources and agricultural yield.

The present study recommends the following measures to minimize the impact of climate change on water resources and crop yield.

- Regions which are most vulnerable to climate change, should adapt strategies to manage climate change risk. Easily and economically feasible techniques and strategies should be formed in the countries falling under this category. The strategies will vary from one country to other depending on their hydrological conditions.
- Countries with arid and semi arid regions should be encouraged to invest and finance research projects related to climate change studies and data collection related to water.
- Improvement in irrigation infrastructure should be carried out in all vulnerable countries to cope with climate change risks. Farmers should be trained for alternative livelihoods in areas expected of severe impact of climate change.
- Programs related to groundwater development such as rainwater harvesting, watershed management, formation of artificial recharge facilities with wastewater and rainwater should be implemented on a large scale in all vulnerable areas.
- Artificial recharge of groundwater with partially treated wastewater can be an added advantage to reuse policies of water in all arid and semi-arid countries. This technique has proven advantages over the direct application of treated wastewater.

- Agricultural sector requires large amounts of water, which will definitely affect water availability for domestic and industrial requirements in future. Therefore more emphasis should be given to watershed development and upgrading rain fed agriculture through rainwater harvesting and artificial recharge systems.
- Multi-disciplinary approach and advanced techniques should be adopted in the evaluation of vulnerability of different water resources and their associated systems that add to global warming.
- Approximately with 90 percent certainty IPCC report revealed that emission of greenhouse gases is responsible for current global warming trends. Therefore all possible efforts should be carried out for reducing green-house gas emission—particularly carbon-dioxide released through the burning of fossil fuel.
- Global warming affects the environment of the entire world. Thus for minimizing the impact of climate change on water management public awareness and involvement are needed. Hydrologists can play an important role in increasing the awareness with discussions with students, business groups, corporates and communities through conferences.

ACKNOWLEDGEMENTS

I thank all the faculty members of the Civil Engineering Department, of ITM University, for providing working facilities and also for continuous encouragement.

REFERENCES

1. Adger et al., 2005, W.N.N.W. Arnell Adger, E.L. Tompkins, Successful adaptation to climate change across scales, Global Environ. Change Part A, 15 (2005), pp. 77–86
2. Anthony Nyong. 2005. Abstract in Impacts of Climate Change in the Tropics: The African Experience. Avoiding Dangerous Climate Change (Meteorological Office, Exeter, UK, 2005) (www.stabilisation2005.com).

3. Batino and Waswa, 2011, A. Batino, B.S. Waswa, New challenges and opportunities for integrated soil fertility management in Africa, A. Bationo, B. Waswa, J.M. Okeyo, F. Maina, J. Kihara (Eds.), Innovations as Key to the Green- Revolution in Africa: Exploring the Scientific Facts, Vol. 1Springer, Dordrecht, The Netherlands (2011), pp. 3–20
4. Burney et al., 2010, J.A. Burney, S.J. Davis, D.B. Lobell, Greenhouse gas mitigation by agricultural intensification, Proc. Natl. Acad. Sci. U. S .A., 107 (26) (2010), pp. 12052–12057
5. Chatdarang, 2001, Chatdarang, V. 2001. Artificial recharge for conjunctive use in irrigation: The San Joaquin valley, California: (M. Tech Thesis); Massachusetts Institute of Technology, June 2001, USA.
6. Chen and Ravallion, 2007, Chen, S.; Ravallion, M. 2007. Absolute poverty measures for the developing world, 1981–2004. World Bank, Development Research Group, Poverty Team.
7. CSAG, 2002. Assessments of Impacts and Adaptations to Climate Change: Africa Climate Change Scenarios: Vol. 1 (2002) (www.csag.uct.ac.za).
8. de Wit and Stankiewicz, 2006, Maarten de Wit, Jacek Stankiewicz, Changes in surface water supply across Africa with predicted climate change, Science, 311 (2006), pp. 1917–1921
9. DHAN foundation, 2002, DHAN foundation, 2002. Revisiting tanks in India, National Seminar on conservation and development of Tanks. New Delhi India.
10. Evenson and Gollin, 2003, R.E. Evenson, D. Gollin, Assessing the impact of the green revolution, 1960 to 2000, Science, 300 (2003), pp. 758–762
11. Fadlelmawla et al., 1999, Fadlelmawla A, M. Abdel-Halen and Vissers, M. 1999. Preliminary plans for artificial recharge with exclaimed wastewater in Egypt; 9th Biennial Symposium on Artificial recharge of groundwater, Tempe, AZ, June 10–12.
12. FAO, 2009, FAO, The State of Food Security in the World, Food and Agriculture Organization of the United Nations, Rome (2009)
13. FAO, 2012, FAO. 2012. Wastewater treatment and use in agriculture. http://www.fao.org/docrep/T0551E/t0551e07.htm.

14. Gosain et al., 2006, A.K. Gosain, Sandhya Rao, Debajit Basuray, Climate change impact assessment on hydrology of Indian river basins, Curr. Sci., 90 (3) (2006), pp. 346–353
15. Gupta and Deshpande, 2004, S.K. Gupta, R.D. Deshpande, Water for India in 2050: first-order assessment of available options, Rev. Pap. Curr. Sci., 86 (9) (2004), pp. 1216–1224
16. Hansen, 2002, J.W. Hansen, Realizing the potential benefits of climate prediction to agriculture: issues, approaches, challenges, Agric. Syst., 74 (3) (2002), pp. 309–330
17. IPCC, 2007, IPCC. 2007. Fresh water resources and their management climate change 2007; Impact, adaptation and vulnerability. Contribution of working group-II to the fourth assessment report of the intergovernmental panel on climate change. Cambridge University Press: Cambridge, UK.
18. IPCC, 2007A, IPCC. 2007A. Climate change, The physical science basis; Working group I contribution to the fourth Assessment report of the Intergovernmental Panel on Climate change; Summary for policymakers, Technical Summary and frequently asked questions.
19. IPCC, 2008, IPCC. 2008. Climate change and water, Intergovernmental panel on climate change technical report IV. June 2008.
20. Jarvis et al., 2010, Jarvis, A., Ramirez, J., Anderson, B., Leibing, C., Aggarwal, P. 2010. Scenarios of climate change within the context of agriculture. In: Climate Change and Crop Production. CAB International.
21. Jones et al., 2003, J.W. Jones, G. Hoogenboom, C.H. Porter, K.J. Boote, W.D. Batchelor, L.A. Hunt, P.W. Wilkens, U. Singh, A.J. Gijsman, J.T. Ritchie, DSSAT cropping system model, Eur. J. Agron., 18 (2003), pp. 235–265
22. Kahinda and Taigbenu, 2011, J.M. Kahinda, A.E. Taigbenu, Rainwater harvesting in South Africa: challenges and opportunities, Phys. Chem. Earth, 36 (2011), pp. 968–976
23. Kundzewicz and Döll, 2009, Z.W. Kundzewicz, P. Döll, Will groundwater ease freshwater stress under climate change? Hydrol. Sci. J. –J. Des. Sci. Hydrol., 54 (2009), pp. 665–675

24. Kundzewicz et al., 2007, Z.W. Kundzewicz, L.J. Mata, N.W. Arnell, P. Döll, P. Kabat, B. Jiménez, *et al.* Freshwater resources and their management M.L. Parry, O.F. Canziani, J.P. Paluticof, P.J. van der Linden, C.E. Hanson (Eds.), Climate Change 2007: Impacts, Adaptation, and Vulnerability. Contributions of the Working Group II to the Fourth Assessment Report of the Intergovernmental Panel on Climate Change, Cambridge University Press, Cambridge, UK (2007), pp. 173–210
25. Kurukulasuriya and Mendelson, 2007, Kurukulasuriya, P., and Mendelsohn, R. 2007. A Ricardian Analysis of the Impact of Climate Change on African Crop Land, Policy Research Working Paper (4305), The World Bank, Washington DC.
26. Kurylyk and MacQuarrie, 2013, Barret L. Kurylyk, Kerry T.B. MacQuarrie, The uncertainty associated with estimating future groundwater recharge: a summary of recent research and an example from a small unconfined aquifer in a northern humid-continental climate, J. Hydrol., 492 (2013), pp. 244–253
27. Mapfumo et al., 2013, P. Mapfumo, Adjei-Nsiah Samuel, Florence Mtambanengwe, Regis Chikowo, KenE. Giller, Participatory action research (PAR) as an entry point for supporting climate change adaptation by smallholder farmers in Africa, Environ. Dev., 5 (2013), pp. 6–22
28. Misra, 2013, A.K. Misra, Global climatic variations and its impact on Indian hydrological conditions, Int. J. Eng. Sci. Res. Technol., 2 (2) (2013), pp. 148–158
29. Misra et al., 2013, A.K. Misra, Manav Wadhwa, Ankur Shivhare, Anupriya Gupta, Nikita, Design and testing of artificial recharge structures equipped with geosynthetic materials for arid and semi arid areas, Adv. Water Resour. Prot., 1 (4) (2013), pp. 53–57
30. MOWR, 1999, MOWR. 1999. Integrated water resources development-a plan for action. Report of the national Commission for integrated Water Resources Development plan; Ministry of Water Resources, Government of India, Vol. 1, pp. 515.
31. Munang et al., 2013, Munang R, Johnson N. Nkem, ZhenHan. 2013. Using data digitalization to inform climate change adaptation policy: Informing the future using the present; Weather and Climate Extremes; http://dx.doi.org/10.1016/j.wace.2013.07.001i.

32. Nellor et al., 1985, M.H. Nellor, R.B. Baird, J.R. Smyth, Health effects of indirect potable water reuse, J. Am. Water Works Assoc., 77 (1985), pp. 88–96

33. Nelson et al., 2007, Nelson, B. Monty Schmitt, Ronnie Cohen, Noushin Ketabi. 2007. Water Management Strategies to Weather the Effects of Global Warming; NRDC reports Department, 40 West 20th Street, New York, NY 10011, USA.

34. O'Hare et al., 1986, O'Hare, M.P., Fairchild, D.M., Hajali, P.A., Canter, L.W. 1986. Artificial recharge of Groundwater. Proceedings of the second International Symposium on Artificial Recharge of Groundwater.

35. RIGW, 1999, RIGW/IWACO. 1999. Artificial recharge pilot projects at El-Bustan area, Egypt. Final report; Recharge institute for groundwater, Egypt.

36. Sakthivadivel, 2007, Sakthivadivel Ramaswamy. 2007. The Groundwater Recharge Movement in India; www.iwmi.cgiar.org/.../Ground_Water/.../Giordano_1845931726-Chapter10.pdf.

37. Solomon et al., 2007, S. Solomon, D. Qin, M. Manning, Z. Chen, M. Marquis, K. Averyt, *et al.* Climate Change 2007: The Physical Science Basis Contributions of Working Group I to the Fourth Assessment Report of the Intergovernmental Panel on Climate ChangeCambridge University Press, Cambridge (2007)

38. Thornton et al., 2011, P.K. Thornton, P.G. Jones, P.J. Ericksen, A.J. Challinor, Agriculture and food systems in sub-Saharan Africa in a 4 °C+ world Philos. Trans. R. Soc. A: Math. Phys. Eng. Sci., 369 (1934) (2011), pp. 117–136

39. UNEP, 2004, UNEP. 2004. http://www.unep.or.jp/ietc/publications/techpublications/TechPub-8e/recharge.asp.

40. UNESCO, 2004, UNESCO. 2004. The water resources in the OSS (Observatory of the Sahara and the Sahel) countries; evaluation, use and management; International Hydrological Programme. Published in 2004; 7, place de Fontenoy, 75352, Paris 07 SP.

41. Vermeulen et al., 2012, S.J. Vermeulen, P.K. Aggarwal, A. Ainslie, C. Angelone, B.M. Campbell, A.J. Challinor, J.W. Hansen, J.S.I. Ingram, A. Jarvis, P. Kristjanson, C. Lau, G.C. Nelson, P.K. Thornton, E. Wollenberg

Chapter 9

Burnt Sugarcane Harvesting – Cardiovascular Effects on a Group of Healthy Workers, Brazil

Cristiane Maria Galvão Barbosa[1,2], Mário Terra-Filho[1], André Luis Pereira de Albuquerque[1], Dante Di Giorgi[3], Cesar Grupi[4], Carlos Eduardo Negrão[5], Maria Urbana Pinto Brandão Rondon[5], Daniel Godoy Martinez[5], Tânia Marcourakis[6], Fabiana Almeida dos Santos[6], Alfésio Luís Ferreira Braga[7,8], Dirce Maria Trevisan Zanetta[9], Ubiratan de Paula Santos[1*]

[1]Pulmonary Division - Heart Institute(InCor), Hospital das Clínicas da Faculdade de Medicina da Universidade de São Paulo, São Paulo, São Paulo, Brazil,

[2]FUNDACENTRO, São Paulo, São Paulo, Brazil,

[3]Hypertension Unit, Heart Institute(InCor), Hospital das Clínicas da Faculdade de Medicina da Universidade de São Paulo, São Paulo, São Paulo, Brazil,

[4]Electrocardiology Unit, Heart Institute(InCor), Hospital das Clı´nicas da Faculdade de Medicina da Universidade de Sa˜o Paulo, Sa˜o Paulo, Sa˜o Paulo, Brazil,

[5]Unit of Cardiovascular Rehabilitation and Exercise Physiology, Heart Institute (InCor), Hospital das Clı´nicas da Faculdade de Medicina da Universidade de Sa˜o Paulo, Sa˜o Paulo, Sa˜o Paulo, Brazil,

[6]Department of Clinical and Toxicological Analyses, University of Sa˜o Paulo Pharmacological Sciences School, Sa˜o Paulo, Brazil,

[7]Environmental Epidemiology Study Group, Laboratory of Experimental Air Pollution, Department of Pathology, Faculdade de Medicina da Universidade de Sa˜o Paulo, Sa˜o Paulo, Sa˜o Paulo, Brazil,

[8]Environmental Exposure and Risk Assessment Group, Catholic University of Santos, Santos, Sa˜o Paulo, Brazil,

[9]Department of Epidemiology, University of Sa˜o Paulo School Public Health, Sa˜o Paulo, Brazil

INTRODUCTION

Brazil is the world's largest producer of sugar and ethanol from sugarcane, with 570 million tons in 2007/2008 harvest [1]. Although industrial harvesting processes utilize technological methods, manual harvesting is still the predominant method of harvesting sugarcane, and it employs nearly 500,000 workers throughout the country. This is a seasonal activity. For seven months per year, to receive an average monthly wage of US$ 700.00 a sugar cane worker must cut approximately 10 tons of sugarcane daily, in journeys of eight hour and twenty minutes, six days a week, under high temperatures in the fields, due to the climate and the heat from burning sugarcane, and receiving inappropriate reposition of water and electrolytes. Moreover, they are exposed to pollutants released during the cutting of burnt sugarcane.

Exposure to air pollution is associated with increased cardio-respiratory morbimortality, and most studies on the subject are related to urban pollution (industrial/vehicular origin) [2], [3], [4],[5]. The studies on outdoor air pollution caused by biomass burning have focused more on the respiratory effects [6], [7] than on the cardiovascular effects [8], [9]. Study carried out with sugar cane workers found in their urine levels of 1-hydroxipirene, an exposure markers for polycyclic aromatic hydrocarbons, 10 times higher during the harvest season [10].

There are currently no published studies on the cardiovascular risks associated with manual harvesting of sugarcane, which combines physical and thermal overload as well as exposure to pollutants under conditions that are present in countries such as Brazil, India, Philippines, Latin and Central America countries. It has been reported sugar cane workers' diseases and sudden deaths in the last decade [11], [12].

Our objective was to evaluate the occurrence of cardiovascular effects and the possible mechanisms involved in these events associated with the harvesting of burnt sugarcane.

METHODS

Study Population and Period

This is an observational study with repeated measures conducted on 28workers at a sugar and ethanol mill. All participants were Caucasian, male, healthy, between 18 and 50 years of age and had no clinical history or use of medications for cardiopulmonary disease.

The study participants were evaluated at two periods: at the end of burnt sugarcane harvest (October–November 2007) and at the end of period when burnt sugarcane was not being harvested (March–April 2008), when the cutters performed cleanup and planting of unburned sugarcane. As they did not earn by productivity in this period, activity was physically less intense. These time points are henceforth referred to as the harvest and non-harvest periods, respectively.

The Research Ethics Committee of the University of São Paulo Medical School approved the study and all participants signed consent forms.

Examination Procedures

Because there are many risk factors as physical and thermal overload as well as exposure to air pollutants, we decide to use a wide range of effect indicators. These markers have been used in exercise, environmental, and occupational health studies. During both periods, workers were divided into groups of five or six. After working all week, they were brought from the countryside to the Heart Institute in São Paulo city, where they underwent several examinations over five consecutive days. Evaluations were conducted sequentially to avoid any changes between exams (Figure 1).

- The participants answered a questionnaire that was developed for the study and involved the following: data regarding the work during both periods, work time, prior occupational exposures, smoking, and the presence of general and respiratory symptoms. Anthropometric measurements were taken at this moment.
- Blood markers: fibrinogen, thrombin time (TT), prothrombin time (PT), platelet count, creatine kinase (CK), lactate dehydrogenase (LDH), lipid profile, serum calcium, serum sodium and C-reactive protein (CRP) by high sensitive immunology assay (Dade Behring Marburg GmbH, Germany), erythrocyte glutathione peroxidase (GPx), Glutathione-S-Transferase (GST) and glutathione reductase (GR) measured by spectrophotometry [13], [14] and plasma malondialdehyde (MDA) measured by high-performance liquid chromatography [15].
- Twenty-four-hour ambulatory blood pressure monitoring (ABPM-24 hour) was performed using a Spacelabs-90207 monitor (Spacelabs Medical Inc., USA). Measurements were taken every 10-minutes during the day (5a.m. to 10 p.m.) and every 20-minutes at night (10 p.m. to 5 a.m.), according to standards that have been defined and used with our services [16].
- Twenty-four-hour electrocardiography monitoring (ECG-24 hour): was performed according to a method that was previously used in another study by our group [17]. Records were analyzed to obtain heart rate variability (HRV) indicators, including the standard deviation of normal RR intervals (SDNN), the standard deviation of sequential five-minute RR interval means (SDANN), and the root mean square of differences between NN adjacent intervals (rMSSD).
- Cardiopulmonary exercise testing (CPET): a ramp symptom-limited CPET was performed on a cycle (Corival, The Netherlands), consisting in a 2-min period of rest, a 2-min period of warm-up (unloaded pedaling) followed by an incremental work-rate

period (increase of 20 W/min) [18]. Oxygen saturation (SpO_2) by pulse oximetry (NONIN-ONYX, model 9500, Plymouth, MN, USA) and electrocardiography (Welch Allyn CardioPerfect, Inc, NY) were monitored continuously. The following variables were recorded breath-by-breath ($CardiO_2System$, MGC): oxygen consumption (VO_2), minute-ventilation (V_E), carbon dioxide output (VCO_2), tidal volume (V_T), respiratory rate, respiratory exchange rate (RER) and heart rate (HR).

- Muscle sympathetic nerve activity (MSNA) was directly measured from the peroneal nerve using the technique of microneurography, as described elsewhere [19]. Multiunit postganglionic muscle sympathetic nerve recordings were made using a tungsten microelectrode (tip-diameter 5–15 μm). Muscle sympathetic bursts were identified by visual inspection and were expressed as burst per 100 heartbeats.
- Forearm blood flow (FBF) was measured by venous occlusion plethysmography (Hokanson, Bellevue, WA, USA), according to methods that have been described elsewhere [19]. FBF (mL/min/100 mL of tissue) was determined based on a minimum of four separate readings. Forearm vascular conductance (units) was calculated by dividing the forearm blood flow by the mean blood pressure (oscillometrically measured).

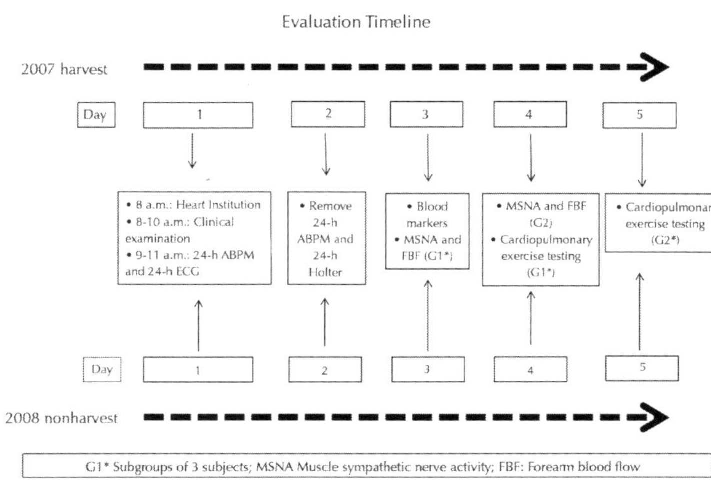

Figure 1: Flow Chart - Evaluation sequence: five groups of six participants.

Work and Environment Assessment

During both periods, we recorded the concentration of Particulate Matter (PM) with a diameter of 2.5 ($PM_{2.5}$) using a DustTrak Aerosol Monitor, model-8520 (TSI-Inc., MN, USA) that was calibrated and adjusted prior to the measurements with a flow rate of 1.7 l/min [20]. On the same days and locations, the temperature (°C) and relative air humidity (%) were measured using a digital thermohygrometer (Dataloger-TFA, 3030.15, Germany). To record environmental variables, an activity cycle was sought to best represent the typical activities of the workers (Figure 2).

Figure 2: Worker cutting burnt sugarcane.

During the harvest period, activities began approximately 4–6 hours after the end of burning. PM and climate variables were measured in the sugarcane field at three 6-hour periods for three consecutive days during the harvest period and during the cutting and weeding of non-burnt sugarcane in the non-harvest period, when there is exposure to resuspended soil particulate matter.

To estimate the level of heat exposure to which the workers were subjected, we obtained the Wet Bulb Globe Temperature (WBGT) Index, a heat exposure indicator [21]. We used a Thermal Stress Meter, model-500 by Quest. The equipment was placed near the worker and readings were made during one workday, according to ACGIH recommendations [22].

Statistical Analysis

Categorical variables are presented as absolute numbers and percentages, whereas continuous variables are given as the mean±standard deviation (SD) or as the median and the interquartile range (IQR). Descriptive analysis were performed for the study variables and the results obtained during the harvest and non-harvest periods were compared by statistical tests for repeated measurements (paired t-test, Wilcoxon rank-test or McNemar's test, as appropriate). Differences of $PM_{2.5}$ and climate variables measurements, between harvest and non-harvest periods, were tested using Mann-Whitney U test.

When the p-value of the differences in measurements for both periods evaluated was$\leq$0.10, linear regression analysis was carried out using a generalized estimating equation (GEE) with robust standard error estimators to evaluate the effects of harvesting work. Adjustments were made for age, body mass index (BMI) and smoking (nonsmoking as a reference), assuming equal correlation of the measurements for each subject (exchangeable correlation). Socioeconomic status was quite similar among the participants and was not included in the analysis. The effects of HRV and MSNA on blood pressure (BP) were evaluated controlling for harvest, age, BMI and smoking (*BP~HRV or MSNA+ harvesting+age+smoking+BMI*). The GEE function was obtained from StatLib (http://lib.stat.cmu.edu/) and the analysis was performed on S-Plus® 8.0 for Windows Statistics software, Insightful Corp., Seattle, WA.

RESULTS

Particulate Matter, Climate and Heat Exposure

Table 1 shows data for $PM_{2,5}$ concentrations, temperature and relative air humidity (RAH) at the sugarcane fields during the two periods. The $PM_{2.5}$ concentration was higher during the harvest period. There was no significant difference in temperature, and RAH was significantly higher during the non-harvest period. WBGT varied between 18.1 and 28.4°C, with the highest value observed between 11a.m.and 12p.m.

Table 1: Descriptive analyses of fine particles ($PM_{2.5}$), ambient temperature and relative humidity during harvest and non-harvest periods at the sugarcane field

Variables	Harvest	Non-harvest	p-value[a]
	Median (IQR[b])	Median (IQR)	
$PM_{2.5}$ ($\mu g/m^3$)	87.0(70.0-100,0)	50.0(40.5-61.5)	<0.001
Ambient Temperature (°C)	28.1(25.6-33.0)	28.2(25.6-31.8)	0.500
Relative Humidity (%)	49.0(40.0-59.0)	65.0(61.0-72.8)	<0.001

a: Mann-Whitney test;

b: interquartile range.

Individual Evaluations

The mean worker age was 31±6.3 years (range: 21–45 years), and the average length of time spent employed in sugarcane harvesting was 9.8±8.4 years. Nineteen workers (68%) were never-smokers and nine (32%) were light smokers (7±4.2 pack-years). All workers had less than eight years of schooling, earned between 500 and 800 US dollars per month, according to the amount of cut cane. The housing conditions were similar, living on the outskirts of small towns in the region. None had the automotive vehicle itself.

As the similar cultural habits and socioeconomic conditions, these workers have the same food habits. They carry to work on the field the meal prepared the day before and stored in bowls partially saving thermal food temperature. However, the fact that there is loss of heat by the time of the meal gave this group the nickname of *"cold-meal workers"*. Usually they take rice and beans, potatoes and some animal protein in the form of egg, meat, or chicken

History of previous chronic disease was absent in all participants and none presented infectious or traumatic events when evaluated.

Table 2 presents the general data for the participants in both periods. The workers showed significantly lower body weight during the harvest period as well as smaller abdominal circumference and BMI. The serum HDL presented higher levels during harvest when compared to non-harvest period. On the other hand, diastolic and, in a lesser extent, the systolic blood pressure were higher during harvest period.

Table 2: General Characteristics of participants during harvest and non-harvest periods in 28 participants

Variables	Harvest	Non-harvest	p-value
BMI[a][kg/m^2 (mean ± SD[b])]	22.6 ± 2.7	23.4±2.9	<0.001[c]
Weight [kg (median; IQR[d])]	64.5 (61.0-69.5)	67.0 (2.0-73.5)	<0.001[e]
Abdcirc[f][cm (median; IQR))	80.0 (75.0-84.0)	83.0 (78.0-87.5)	0.002[e]
Total Cholesterol [mg/dL(mean ± SD)]	159.9 ± 38.1	151.2 ± 33.5	0.041[c]
HDL[g] Cholesterol [mg/dL(mean± SD)]	50.0 ± 10.7	42.5 ± 8.2	<0.00[c]
LDL[h] Cholesterol [mg/dL(mean ± SD)]	90.0 ± 26.1	88.8±27.0	0.689[c]
Triglycerides [mg/dL(mean ± SD)]	102.6 ± 64.2	106.7 ± 91.4	0.663[c]
Heart rate [BPM'(mean ± SD]	60.8 ±8.8	58.7 ± 9.1	0.237[c]
SBP[j][mmHg (mean ± SD)]	125 36± 14,81	118.07±17.8	0.046[c]
DBP[k][mmHg (mean ± SD)]	78.00 ± 12.96	70.57±13.88	0.013[c]

a: body mass index;
b: standard deviation;
c: paired t test;
d: interquartile range;
e: Wilcoxon rank test;
f: abdominal circumference;

g:High Density Lipoprotein;
h: Low Density Lipoprotein;
i: beats per minute;
j: systolic blood pressure;
k: diastolic blood pressure.

Sixteen workers (51.7%) reported nasal itching and rhinorrhea during the harvest period, versus four (14.3%) in the non-harvest period ($p<0.01$); ten workers (35.7%) mentioned cramping during the harvest period, compared with three (10.7%) in the non-harvest period ($p = 0.02$); seven workers (25.0%) reported a dry cough during the harvest period, versus only one (3.6%) in the non-harvest period ($p = 0.02$).

During the harvest period, the workers reported a daily water intake of 5–10 liters. The average weight of the sugarcane that was cut daily, as reported by the workers, was 11 tons (range: 7–14), differently from non-harvest period, when workers receive a pre-defined payment by month and, therefore, the job is less intensive.

Table 3 and Figure 3 shows the blood marker results for the workers during both periods. The CK and LDH levels were higher, while CRP, calcium, TT and PT were significantly lower during the harvest period. Sixteen (57%) workers showed sodium levels below 140 meq/L during the harvest period, versus six (21.5%) in the non-harvest period. Fibrinogen levels were higher in the harvest period, but the differences were not significant.

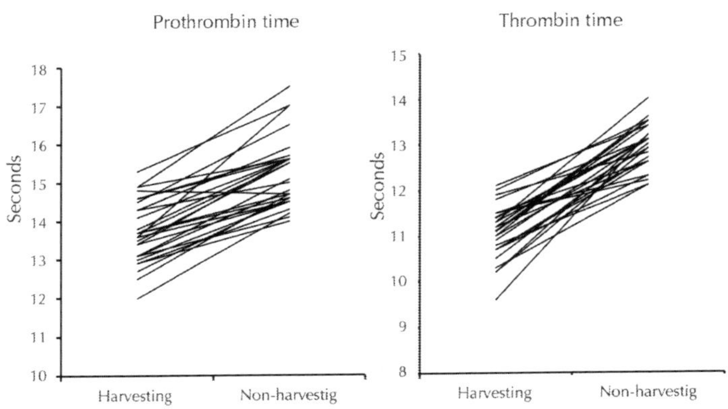

Figure 3: Prothrombin and Thrombin time in harvest and non-harvest periods (n:28).

Table 3: Descriptive analyses of blood markers during harvest and non-harvest periods and the estimated effects of harvest using regressions analysis for repeated measures in 28 participants

Variables	Periods		Univariate Analysis	Multiple Analysis'	
	Harvest	Non-harvest	p-value	RC (95%CI)[b]	p-value
Sodium (mEq/L)	139.6±2.3[c]	141.6±1.6[c]	0.046[d]	-0.86 (-1.84; 0.12)	0.085
Calcium (mg/dl)	9.1 ±0.4[c]	9.3±0.3[c]	0.013[d]	- 0.21(- 0.36; -0.07)	0.004
CK[e](U/L)	136.5 (108.5-216.0)[f]	104.5 (77.5-170.5)[f]	<0.001[g]	39.06 (19.24; 58.87)	<0.001
LDH[h](U/L)	156.6 -± 22.0[c]	148.1 ± 23.9[c]	0.028[d]	9.44 (2.10; 16,78)	0.012
CRP[i](mg/L)	2.5 (1.6-5.3)[f]	4.4 (1.7-15.2)[f]	<0.001[g]	-4.46 (-6,89; -2.02)	<0.001
PT[j](s)	13.7±0.8[c]	15.3±0.9[c]	<0.001[d]	-1.50 (-1.77; -1.23)	<0.001
TT[k](s)	11.1±0.6[c]	13.0 ± 0.5[c]	<0.001 [d]	-1.86 (-2.10; -1.62)	<0.001

a: multiple analysis adjusted for age, body mass index and smoking;
b: RegressionCoefficientand95%Confidence interval;
c: mean and standard deviation;
d: paired t test;
e: creatine kinase;
f: median and interquartile range;
g: Wilcoxon runk test;
h: lactate dehydrogenase;
i: C-Reactive Protein;
j: prothrombin time;
k: thrombin time.

The enzymes GR, GST and GPX activity, as well as MDA levels, were higher in the harvest period and the difference was significant for the GST and GPX (Table 4).

Table 4: Glutathione transferase (GST), glutathione peroxidase (GPX), glutathione reductase (GR), and malondialdehyde (MDA) during harvest and non-harvest periods in 28 participants

Variables	Periods		Paired t test	Multiple analysis[a]	
	Harvest (mean ± SD[b])	Non-harvest (mean±SD)	p-value	RC (95^0/0CI)`	p-value
GST[d](Ug/Hb)	3.38±1.27	3.01 ±1.31	0.019	0.39(0.16; 0.63)	0.01
GPX' (Ug/Hb)	55.06±11.84	39.48±9.45	0.001	15.31(9.81; 20.82)	<0.001
GR[f] (Ug/Hb)	3.05±0.97	2.91 ±1.13	0.401	0.23(-0.08; 0.54)	0.143
MDA[9] (pWdL)	7.50±1.42	6.89±1.01	0.088	6.4(-0.26; 13.06)	0.057

a: multiple analysis adjusted for age, body mass index and smoking;
b: standard deviation;
c: regression coefficient and 95% confidence interval;
d: Glutathione - S- Transferase;
e: Glutathione Peroxidase;
f: Glutathione Reductase;
g: Malondialdehyde.

Blood Pressure and Heart Rate Variability

The 24 hours Systolic and Mean BP were significantly higher during the harvest period (Table 5). HRV indicators tended to be higher in the harvest period, although the difference in multiple regression analysis was significant only for SDANN, the estimate of long-term components of HRV. The frequency domain variables did not present statistically significant variations.

Table 5: Ambulatory Blood Pressure Monitoring (ABPM) and Heart Rate Variability (HRV) during harvest and non-harvest periods in 28 participants

Variables	Periods (meani-SD[a])		Paired t test	Multiple analysis[b]	
	Harvest	Non-harvest	p-value	RC (95%CI)[c]	p-value
24 h ABPM[d]					
SBP[e] (mmHg)	120.1 ± 103	117.0±10.0	0.110	3.69 (0.27; 7.11)	0.034
MBP[f](mmHg)	86.8±9.4	84.4±8.6	0.103	2.95 (0.15; 5.75)	0.039
24 h ECG-HRV[g]					
SDNN[h](ms[i])	187.1±37.7	178.6±40.5	0.030	5.81 (-2.77; 14.39)	0.184
SDANN[j](ms)	161.6±30.4	149.0 ± 34.0	0.006	11.22 (2.40; 20.03)	0.013
RMSSD[k](ms)	47.5±19.1	47.6±19.4	0.957	-2.52 (- 7.19;2.15)	0,290

a: standard deviation;
b: multiple analysis adjusted for age, body mass index and smoking;
c: regression coefficient and 95% confidence interval;
d: Twenty-four-hour Ambulatory blood pressure monitoring;
e: Systolic blood pressure;
f: Mean blood pressure;
g: Twenty-four-hour electrocardiogram-heart rate variability;
h: standard deviation of normal RR intervals;
i: milliseconds;
j: Standard Deviation of Sequential Five-Minute R-R Interval Means;
k: root mean square of successive differences in adjacent NN intervals.

Sympathetic Nerve and Vascular Measures

Among the evaluated individuals, 25 adequately completed these tests during both stages. The values for MSNA and FBF at rest and during handgrip exercise tended to be higher during the harvest period, although the differences were not significant (Table 6).

Table 6: Descriptive analyses of muscle sympathetic nerve activity (MSNA) and forearm blood flow (FBF) at rest and during handgrip exercise in 25 participants

Variable	Harvest mean±SD[a]	Non-harvest mean±SD	p-value[b]
MSNA (bursts/100HB[c])[d]			
Baseline	38.9 ± 11.0	36.1 ±8.6	0.056
1′exercise	38.2± 10.9	36.0 ±8.9	0.339
2′exercise	40.9 ±9.3	39.0 ±8.9	0.440
3′exercise	45.5± 10.1	43.5 ±10.4	0.460
FBF (ml/min/100 ml of tissue)			
Baseline	2.9 ± 1.4	2.8 ±1.2	0.657
1′exercise	3.3 ± 1.5	3.1 ±1.4	0.559
2′exercise	3.3 ± 1.8	3.0 ±1.3	0.356
3′exercise	3.3 ±1 9	3.2 ± 1.4	0.602

a: standard deviation;

b: p-value: Paired T-test;

c: heart beats;

d: during 30% of maximum voluntary contraction.

Cardiopulmonary Exercise Testing

Among the evaluated individuals, 24 adequately completed this test during both stages. The mean duration of exercise was 12 minutes. The peak O_2 uptake (peak VO_2), the O_2 pulse and the peak systolic/diastolic BP were significantly higher during the harvest period (Table 7).Figure 4 shows that the systolic (SBP) and diastolic (DBP) blood pressures were higher in the harvest period during all test stages. DBP showed an 11.1 mmHg increase at the end of exercise during the harvest period ($p<0.001$), versus 5.1 mmHg in the non-harvest period ($p = 0.064$).

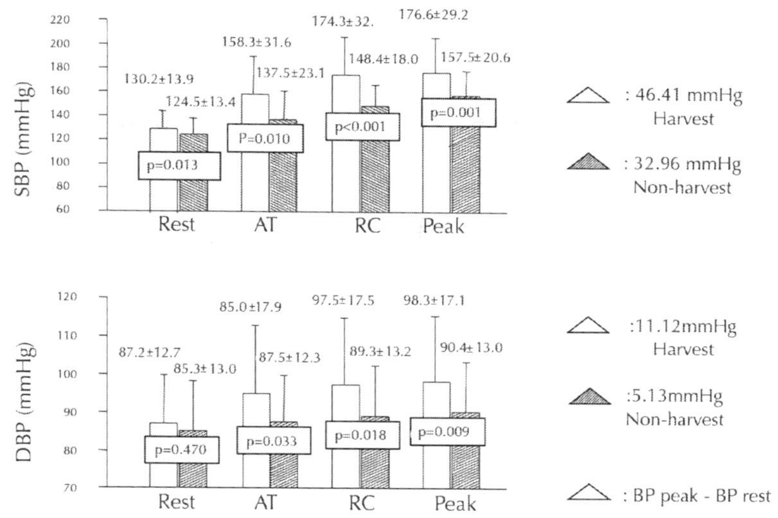

Figure 4: Systolic (SBP) and diastolic (DBP) blood pressure variation during cardiopulmonary exercise testing in each period, mean±SD, (n:24). AT: Anaerobic Threshold; RC: respiratory compensation; Δ = differences between peak and rest in mmHg.

Table 7: Univariate analysis and the estimated effects of harvest using regression models for repeated measures on cardiopulmonary exercise testing in 24 participants

Variables	Periods		Paired t test	Multipleanalysis[a]	
	Harvest[b]	Non-harvest[b]	p-value	RC (95%CI)[c]	p-value
Peak VO_2 (ml/Kg/min)	40.4 ± 6.9	36.5 ± 6,3	0.005	2.86 (0.68; 5.05)	0.010
Peak O_2 (ml/beat)	16.3 ±3.0	14.9±3.2	0.009	1.51 (0.64; 2.38)	0.001
SBP[d]rest(mmHg)	130.2±13.9	124.5 ± 13.5	0.013	6,94 (3.02; 10.86)	0.001
SBPpeak(mmHg)	176.6±29.9	157.6±20.6	0.001	21.59 (11.58; 31.60)	<0.001
DBP[e]rest(mmHg)	87.2 ± 12.7	85.3±13.0	0.470	3.69 (- 1.12;8.50)	0.133
DBP peak (mmHg)	98.3±17.1	90.4±13.0	0.009	10.15 (4.98; 15.32)	<0.001

a: multiple analysis adjusted for age, body mass index and smoking;

b: mean6 standard deviation;

c: regression coefficient and 95% confidence interval;
d: systolic blood pressure;
e: diastolic blood pressure.

Multiple analyses showed that the BP values recorded by ABPM during the harvest period were significantly associated with reductions in HRV and increases in MSNA (Table 8). A 10 ms reduction in SDNN and in rMSSD and a 10 burst/min increase in sympathetic nerve activity were associated with 0.7, 2.2, and 1.8 mmHg rises in 24-hours SBP, respectively.

Table 8: Effects on blood pressure associated with heart rate variability (HRV) and muscle sympathetic nerve activity (10×)

Variable	24 h SBP[a]		24 h DBP[b]		24 h MBP[c]	
	Regression Coefficients (95%CI[d])	p value	Regression Coefficients (95%CI)	p value	Regression Coefficients (95%CI)	p value
HRV[e] (n:28)						
SDNN[f] (ms)	-0.7 (-1.5; 0.03)	0.058	-0.62 (-1.2; -0.1)	0.029	-0.7 (-1.3; -0.1)	0.02
RMSSD[g] (ms)	-2.2 (-4.1; -0.4)	0.019	-1.6 (-3.0; -2.5)	0.021	-2.4 (-3.9; -0.8)	0.009
HF[h](ms)	-5.6 (-9.4; -1.8)	0.004	-3.7 (-6.8; -0.6)	0.021	-4.8 (-8.1; -1.4)	0.005
LF[i] (ms)	-3.7 (-7.7; 0.2)	0.065	-3.0 (-5.9; -0.06)	0.046	-3.5 (-6.8; -0.1)	0.044
LF/HF	28.3 (17.3; 39.3)	<0.001	-7.4 (-26.4; 11.5)	0,519	26.4 (18.8; 33.9)	<0.001
MSNA[j]/MVC[k] (n=25)	1.8 (0.1; 3.5)	0.035	1.4 (0.05; 2.7)	0.049	1.6 (0.8; 2.4)	0.041

DISCUSSION

This study revealed that sugarcane workers are subjected to several risks, such as physical overload in hot conditions and exposure to pollutants. Combined, these conditions, which were observed during the harvest period, are related to changes in cardiovascular and blood markers.

In agreement with other authors [11], the present findings indicated the adverse conditions to which sugarcane harvesters are subjected. Studies of the movements of workers during sugarcane harvesting have estimated that cutters bend their backs approximately 4,000 times and make close to 3,800 machete strikes during an 8-hour work day [23].

There are currently few published studies evaluating sugarcane workers [10], [20]. The existing studies have evaluated pollutants in the area where the sugarcane is burned and in nearby cities [24]. It is also likely that the PM associated with sugarcane burning presents distinct characteristics during the different periods, as can be observed by the blackish color of the workers' clothes in the harvest period. During burning, (Figure 2) we observed suspended matter present in the sugar cane and on the surface of the top layer of soil, resulting from the burning of waste straw that is released by machete strikes, whereas during the non-burning period the suspended PM contains mainly particles suspended from soil.

The WBTG value was high, reaching 28.43°C, similar to the findings of another study on sugarcane harvesting [23] which that registered 27.9°C. These values surpass the recommended limits for continuous labor [21], suggesting that heat stress can occur over the working months, which demonstrates the need for breaks during the workday. Several studies have related the effect of variations of temperature in the cardiovascular system [24], a condition experienced by workers on a daily basis. In this study, the WBTG varied from 18°C to 28°C, what could explain the reports of hospitalization and deaths that have been described in the sugar cane plantation [20].

Higher levels of CK and LDH are compatible with labor under conditions of physical overload and hydro-electrolyte imbalance, as evidenced in several studies [25], [26]. LDH and CK are biomarkers of muscular damage and may increase during situations of intense exercise, in which cell membranes become more permeable and enzymes are released into the interstitial matrix and reabsorbed via the lymphatic system to enter the bloodstream [26].The observation that CK remained high, even in blood samples measured 60 hours after exercise suggests a state of persistent hyperCKemia, much like that which occurs in athletes [26]. Performing exhaustive exercise in an excessively warm environment increases the risk of muscle lesion[26].

Although serum sodium is a less sensitive marker, the lower sodium and calcium levels observed during the harvest period can arise due to possible hydro-electrolyte disturbances associated with intense sweating and the intake of large quantities of water, which could explain the higher frequency of reported cramping in this period [25], [27]. Another study [28]reported the occurrence of hyponatremia among military personnel during intensive training, due to excessive water intake.

The changes observed in TT and PT may be related to exposure to pollutants, as suggested in other studies [29] and likely indicate increased blood viscosity. A study [30] involving 38 individuals showed that metals present in the water-soluble fraction of air pollution particles have an important role in decreasing the whole-blood coagulation time. These findings [30] and those of a recently published study [31] analyzing the composition and effects of PM arising from vehicular emissions and sugarcane burning found a higher concentration of metals (Ni, Fe, Zn, Mn) in the latter, supporting the findings of short PT and TT observed in our study.

The lower CRP values during the harvest period do not concur with the findings of several studies on pollution [32], [33]. However, other studies did not show an association between exposure to pollutants and CRP [34], [35]. Our findings may be due to the positive effect of increased physical activity during the harvest period. Studies performed on athletes have shown a reduction in CRP after intensive physical training [36], [37]. The same was observed in the lipid profile. The only statistically significant variation was observed for HDL cholesterol, which increased during the harvest period, probably due to the more intense physical activity performed during this period.

The increased activities of antioxidant enzymes during the harvest period may be associated with a greater stimulus by oxidant radicals produced by exercise [38], [39] and inhalation of pollutants [40], [41]. The increased levels of MDA, a lipid peroxidation marker [39], during harvest (Table 4), although not statistically significant ($p = 0.057$), may suggest cell damage even with increases of GST and GPX enzymes, indicating an upper regulation of antioxidant defense mechanism [42]. This suggests that both processes may be concomitant. Poorly planned physical activity that is excessive in intensity and pace and is associated with exhaustion – similar to what occurs during the harvest period – may also induce the development of oxidative stress [38], [39].

The reduction of SDANN observed in this study, which has matched the ultralow frequency variability, suggests negative influence on cardiac autonomic balance, a known cardiovascular risk factor [43], it has been associated with air pollution exposures [44], [45], [46].

Higher O_2 pulse and peak VO_2 levels observed in CPET during the harvest period may be associated with an improvement in the physical conditioning of workers during that period [18]. Peak VO_2 is an index used to measure performance in athletes from whom high VO_2max values are expected. One study on sugarcane harvesters in Colombia showed a VO_2max of 42 ml/kg/min, which is close to the value found in our study [47].The values found in this study were similar to those found in sedentary individuals [48] but lower than those in endurance athletes and higher than those in strength athletes [49]. This results are compatible with the activities of sugar cane workers whose labor requires a combination of physical strength and endurance to cut around tens tons per day of sugar cane and carry them from the field to the trucks.

Blood pressure levels raised more during the harvest period, both at rest and during exercise, which may be associated with several mechanisms. Experimental studies in animals [50] and humans [35] as well as epidemiological studies [17], [51], [52] have shown an association between exposure to vehicular and biomass air pollutants and both acute and chronic elevated BP [53]. The main hypothesis about the effect of air pollution involves oxidative stress and pulmonary inflammation [40] inducing systemic inflammation that could cause vascular dysfunction, endothelial dysfunction, vasoconstriction and cardiac remodeling [50], [54]. An imbalance of the autonomic nervous system (ANS) can occur due to the stimulation of intrapulmonary receptors of the ANS. This imbalance can induce changes in the cardiovascular system, resulting in adrenergic vasoconstriction, increased cardiac output and activation of the renin-angiotensin system, leading to endothelial dysfunction and vasoconstriction [54].

By revealing an increase in ABPM-measured blood pressure, an abnormal DBP response during exercise, and the presence of lipid peroxidation days after the end of exposure, our data suggest a subacute effect associated with systemic stress and inflammation [54]. These effects could account for the ANS imbalance evidenced by the relationship between reduced HRV and increased MSNA and higher BP (Table8), as suggested in a recent study [35].

Although the effects of pollution on the sodium balance in the kidneys are unknown, a possible hydro-electrolyte imbalance resulting from strenuous labor under thermal overload conditions may be an additional factor acting on the renin-angiotensin system and thus contributing to higher BP [55] possibly leading to synergism between PM exposure and high temperatures, as suggested in some studies [56], [57]. Another possible explanation could be a chronic rise in vasopressin levels resulting from the rhythm of water intake after intense sweating – in other words, thirst, which happens daily during work. This increase may be associated with higher BP, as suggested in other studies [58].

The persistently inadequate response of DBP during exercise, even during the non-harvest period, suggests the presence of subacute and chronic changes in the cardiovascular system.

This investigation was delineated to evaluate the possible causes associated to sudden deaths that are observed among sugar cane workers. Besides strenuous outdoor work under unsatisfactory feeding and hydration conditions, they are exposed to particulate matter that could increase adverse effects. Different from other study [59] that showed a small systemic proinflammatory response induced only by intermittent moderate-intense exercise, we observed effects on blood pressure and coagulation markers inversely related with regular physical exercise [60]. Increases of blood pressure (mean, systolic, and diastolic, the last during exercise test) and shortening of thrombin and prothrombin times during harvest period could be attributed to pollution exposure.

In summary, it was not possible to evaluate the isolated impact of each risk factor (air pollution, heat and exertion) on the alterations found (changes in coagulation, oxidative stress and blood pressure). However, these alterations suggest an important impact of the work environment on health. As such, these changes can account for the reports of morbimortality among these workers, which affects the most susceptible and/or those who put forth the most effort to achieve a higher income.

The great number of agricultural workers – as well as other workers, such as miners – facing similar conditions around the world makes this problem a public health issue that deserves more attention from different public administration sectors.

Limitations

There are limitations in our study. One result from our inability to better describe individual worker exposure during a longer environmental evaluation period. However, the great homogeneity of the work involved in cutting sugarcane and the minor changes in meteorological variables suggest that the variation in $PM_{2.5}$ is not relevant, considering that our analyses compared the effects of harvest and non-harvest periods. The existence of a control group could have helped the comparison of certain outcomes, but the challenges in creating a group to compare several conditions (temperature, strenuous exercise, pollution and socioeconomic condition) could not be overcome. The fact that we used repeated measurements to compare the effects of different conditions on the same subjects minimizes this issue.

CONCLUSIONS

Sugarcane work during the harvesting period exposes workers to higher levels of particulate matter, thermal overload, and intense physical exertion, inducing muscle lesion, changes in blood coagulation and in heart rate variability, systemic oxidative stress, and high blood pressure. The autonomic imbalance seems to be one of the mechanisms involved in blood pressure changes.

REFERENCES

1. FAO Food and Agriculture Organization of the United Nations. Available:http://faostat.fao.org/site/339/default.aspx. Accessed on 2012 Apr 30.
2. World Health Organization (2006) WHO Air quality guidelines for particulate matter, ozone, nitrogen dioxide and sulfur dioxide. Global update 2005. Summary of risk assessment. Geneva.
3. Krewski D, Jerrett M, Burnett RT, Ma R, Hughes E, et al. (2009) Extended follow-up and spatial analysis of the American Cancer Society study linking particulate air pollution and mortality. Res Rep Health Eff Inst 5–114; discussion 115–136.

4. Brunekreef B, Beelen R, Hoek G, Schouten L, Bausch-Goldbohm S, et al. (2009) Effects of long-term exposure to traffic-related air pollution on respiratory and cardiovascular mortality in the Netherlands: the NLCS-AIR study. Res Rep Health Eff Inst 5–71; discussion 73–89. doi: 10.1289/ehp.10767
5. Feng J, Yang W (2012) Effects of particulate air pollution on cardiovascular health: a population health risk assessment. PLoS One 7: e33385. doi: 10.1371/journal.pone.0033385
6. Naeher LP, Brauer M, Lipsett M, Zelikoff JT, Simpson CD, et al. (2007) Woodsmoke health effects: a review. Inhal Toxicol 19: 67–106. doi: 10.1080/08958370600985875
7. Cançado JED, Saldiva PH, Pereira LAA, Lara LBSS, Artaxo P, et al. (2006) The impact of sugar cane-burning emissions on the respiratory system of children and the elderly. Environ Health Perspect 114: 725–729. doi: 10.1289/ehp.8485
8. Arbex MA, Saldiva PH, Pereira LAA, Braga ALF (2010) Impact of outdoor biomass air pollution on hypertension hospital admissions. J Epidemiolo Community Health 64: 573–579. doi: 10.1136/jech.2009.094342
9. Henderson SB, Brauer M, Macnab YC, Kennedy SM (2011) Three measures of forest fire smoke exposure and their associations with respiratory and cardiovascular health outcomes in a population-based cohort. Environ Health Perspect 119: 1266–1271. doi: 10.1289/ehp.1002288
10. Bosso RMV, Amorin LMF, Andrade SJ, Rossini A, Marchi MMR, et al. (2006) Effects of genetic polymorphisms CYP1A1, GSTM1 and GSTP1 on urinary 1-hidroxypyrene levels in sugar cane workers. Science of the Total Enviromental 382–390. doi: 10.1016/j.scitotenv.2006.07.025
11. Alves F (2006) Por que morrem os cortadores de cana. Saúde e Sociedade 15: 90–98. doi: 10.1590/s0104-12902006000300008
12. Ribeiro H (2008) Sugar cane burning in Brazil: respiratory health effects. Rev Saúde Pública 42: 1–6.
13. Habig WH, Pabst MJ, Jakoby WB (1974) Glutathione S-transferases the first enzymatic step in mercapturic acid formation. J Biol Chem 22: 7130–7139.

14. Flohé L, Günzler WA (1984) Assays of glutathione peroxidase. Methods Enymol 105: 114–121. doi: 10.1016/s0076-6879(84)05015-1
15. Sim AS, Salonikas C, Naidoo D, Wilcken DE (2003) Improved method for plasma malondialdehyde measurement by high-performance liquid chromatography using methyl malondialdehyde as an internal standard. J Chromatrogr B 785: 337–344. doi: 10.1016/s1570-0232(02)00956-x
16. Brazilian Society of Cardiology, Brazilian Society of Hypertension, and Brazilian society of Nephrology (2005) IV Guideline for ambulatory blood pressure monitoring. IV ABPM/II HBPM. Arquivos Brasileiros de Cardiologia 85: 1–18.
17. Santos UP, Braga ALF, Giorgi DM, Pereira LAA, Gruppi CJ, et al. (2005) Effects of air pollution on blood pressure and heart rate variability: a panel study of the vehicular traffic controllers in the city of São Paulo, Brazil. Eur Heart J 26: 193–200. doi: 10.1093/eurheartj/ehi035
18. ATS ACPP (2003) Statement on cardiolopulmonary exercise testing. Am J Respir Crit Care Med 167: 211–277. doi: 10.1164/rccm.167.2.211
19. Negrão CE, Rondon MUPB, Tinucci T, Alves MJN, Roneda F, et al. (2001) Abnormal neurovascular control during exercise is linked to heart failure severity. Am J Physiol 280: H286–292.
20. Goto DM, Obuti CA, Barbosa CMG, Saldiva PHN, Zanetta DMT, et al. (2011) Effects of biomass burning on nasal mucuciliary clearance and mucus properties after sugarcane harvesting. Environmental Research 111: 664–669. doi: 10.1016/j.envres.2011.03.006
21. ACGIH® (2010) American Conference of Governamental Industrial Hygienist. TLVs®and BEIs® - Threshold Limit values and Biological Exposure Indices.
22. ACGIH (2010) American Conference of Governamental Industrial Hygienist. TLVs®and BEIs® - Threshold Limit values and Biological Exposure Indices.
23. Laat EF, Vilela RAG, Silva AJN, Luz VG (2008) Impact over the working conditions:physical wear of sugar-cane cutters. In: Plataforma Banco Nacional de Desenvolvimento Econômico

e Social (BNDES). Instituto Brasileiro de Análises Sociais e Econômicas (IBASE), eds Impactos da indústria canavieira no Brasil. Rio de Janeiro: IBASE. pp. 36–46.

24. Cançado JED, Braga ALF, Pereira LAA, Arbex MA, Saldiva PH, et al. (2006) Clinical repercussions of exposure to atmospheric pollution. J Bras Pneumol 32: s5–11. doi: 10.1590/s1806-37132006000800003
25. Shieh SD, Lin YF, Lu KC, Li BL, Chu P, et al. (1992) Role of creatine phosphokinase in predicting acute renal failure in hypocalcemic exertional heat stroke. Am J Nephrol 12: 252–258. doi: 10.1159/000168454
26. Brancaccio P, Maffuli N, Limogelli FM (2007) Creatinine kinase monitoring in sport medicine. British Medical Bulletin 81: 287–230. doi: 10.1093/bmb/ldm014
27. Hew-Butler T (2010) Arginine vasopressin, fluid balance and exercise: is exercise-associated hyponatraemia a disorder of arginine vasopressin secretion? Sports Med 40: 459–479. doi: 10.2165/11532070-000000000-00000
28. Garigan TP, Ristedt DE (1999) Death from hyponatremia as result of acute water intoxication in a Amy Basic Trainee. Mil Med 164: 234–238.
29. Baccarelli A, Zanobetti A, Martinelli I, Grillo P, Hou L, et al. (2007) Effects of exposure to air pollution on blood coagulation. J of Thrombosis and hemostasis 5 (2) 252–260. doi: 10.1111/j.1538-7836.2007.02300.x
30. Sangani RG, Soukup JM, Ghio AJ (2010) Metals in air pollution particles decrease whole-blood coagulation time. Inhal Toxicol 22: 621–626. doi: 10.3109/08958371003599037
31. Mazzoli-Rocha F, Magalhães CB, Malm O, Saldiva PH, Zin WA, et al. (2008) Comparative respiratory toxicity of particles produced by traffic and sugar cane burning. Environ Research 108: 35–41. doi: 10.1016/j.envres.2008.05.004
32. Peters A, Frobtich M, Doning A, Immervoll T, Wichmann HE, et al. (2001) Particulate air pollution is associated with an acute phase response in men. Results from the MONICA-Augsburg Study. Eur Heart Journal 22: 1198–1204. doi: 10.1053/euhj.2000.2483

33. Hoffmann B, Moebus S, Dragano N, Stang A, Mohlen-Kamp S, et al. (2009) Chronic residential exposure to particulate matter air pollution and systemic inflammatory markers. Environ Health Perspect 117: 1302–1308. doi: 10.1289/ehp.0800362
34. Forbes LJL, Patel MD, Rudnicka AR, Alicja R, Cook DJ, et al. (2009) Chronic exposure to out door air pollution and marker of systemic inflammation. Epidemiology 20: 245–253. doi: 10.1097/ede.0b013e318190ea3f
35. Brook RD, Urch B, Dovonch JT, Bard RL, Speck M, et al. (2009) Insights into mechanisms and mediators of the effects of air pollution exposure on blood pressure and vascular function in healthy humans. Hypertension 54: 659–667. doi: 10.1161/hypertensionaha.109.130237
36. Anderson J, Jansson JH, Hellsten G, Nilisson TK, Hallmans G, et al. (2009) Effects of heavy endurance physical exercision inflammatory markers in non athletes. Atherosclerosis 10: 1–5. doi: 10.1016/j.atherosclerosis.2009.10.025
37. Lakka TA, Lakka HM, Rankinen T, Leon AS, Rao DC, et al. (2005) Effect of exercise training on plasma levels of C-reative-protein in healthy adults: The HERITAGE Family Study. Eur Heart J 19: 2018–2025. doi: 10.1093/eurheartj/ehi394
38. Tanskanen M, Atalay M, Uusitalo A (2010) Altered oxidative stress in overtrained athletes. J of Sports Sciences 28: 309–317. doi: 10.1080/02640410903473844
39. Marzatico F, Pansarasa O, Bertorelli L, Sonezini L (1997) Blood free radical antioxidant enzymes and lipidic peroxides following long distance and lactacidemic performances on highly trained aerobic and sprint athletes. J Sports Med Phys Fitness 37: 235–239.
40. Romieu I, Castro-Giner F, Kunzli N, Sunyer J (2008) Air pollution, oxidative stress and dietary supplementation: a review. Eur Respir J 31: 179–197. doi: 10.1183/09031936.00128106
41. Li N, Hao M, Phalen RF, Hinds WC, Nel AE (2003) Particulate air pollutants and asthma. A paradigm for the role of oxidative stress in PM-induced adverse health effects. Clinical Imunology 109: 250–265. doi: 10.1016/j.clim.2003.08.006

42. Gomes EC, Silva AN, de Oliveira MR (2012) Oxidants, antioxidants, and the beneficial roles of exercise-induced production of reactive species. Oxid Med Cell Longev 2012: 756132. doi: 10.1155/2012/756132
43. Task Force of Eur Soc Cardiology, North Am Society, Eletrophysiology (1996) Heart rate variability. Satandarts of measurement physiological interpretation and clinical use. Circulation 93: 1043–1065.
44. Pope CA 3rd, Verrier RL, Lovet EG, Larson AC, Raizenne ME, et al. (1999) Heart variability associated with particulate air pollution. Am Heart J 138: 804–807. doi: 10.1016/s0002-8703(99)70014-1
45. Gold DR, Litonjua A, Schwartz J, Lovett E, Larson A, et al. (2000) Ambient pollution and heart rate variability. Circulation 101: 1267–1273. doi: 10.1161/01.cir.101.11.1267
46. Liao D, Duan Y, Whitsel EA, Zheng ZJ, Heiss G, et al. (2004) Association of higher levels of ambient criteria pollutants with impaired cardiac autonomic control: a population-based study. Am J Epidemiol 159: 768–777. doi: 10.1093/aje/kwh109
47. Spurr GB, Maksud MG, Barac-Nieto M (1977) Energy expenditure, productivity and physical work capacity of sugarcane loaders. The Am J of Clinical Nutrition 30: 1740–1746.
48. Herdy AH, Uhlendorf D (2011) [Reference values for cardiopulmonary exercise testing for sedentary and active men and women.]. Arq Bras Cardiol 96: 54–59.
49. Anastasakis A, Kotsiopoulou C, Rigopoulos A, Theopistou A, Protonotarios N, et al. (2005) Similarities in the profile of cardiopulmonary exercise testing between patients with hypertrophic cardiomyopathy and strength athletes. Heart 91: 1477–1478. doi: 10.1136/hrt.2004.055053
50. Yin Z, Yue P, Xu X, Zhong M, Sun Q, et al. (2009) Air pollution and cardiac remodeling: a role for Rho\Rho-Kinase. Am J Physiol Heart Circ Physiol 296: H1540–H1550. doi: 10.1152/ajpheart.01270.2008
51. Delfino RJ, Tjoa T, Gillen DL, Staimer N, Polidori A, et al. (2010) Traffic-related air pollution and blood pressure in elderly subjects with coronary artery disease. Epidemiology 21: 396–404. doi: 10.1097/ede.0b013e3181d5e19b

52. Baumgartner J, Schauer JJ, Ezzati M, Lu L, Cheng C, et al. (2011) Indoor Air Pollution and Blood Pressure in Adult Women Living in Rural China. Environ Health Perspect doi: 10.1289/ehp.1003371

53. Fuks K, Moebus S, Hertel S, Viehmann A, Nonnemacher M, et al. (2011) Long-Term Urban Particulate Air Pollution, Traffic Noise and Arterial Blood Pressure. Environ Health Perspect doi: 10.1289/ehp.1103564

54. Brook RD, Rajagopalan S, Pope-CA r, Brook JR, Bhatnagar A, et al. (2010) Particulate matter and air pollution and cardiovascular disease: An update to the scientific statement from the American Heart Association. Circulation 121: 2331–2378. doi: 10.1161/cir.0b013e3181dbece1

55. Alpérovitch A, Lacombe JM, Hanon O, Dartigues JF, Ritchie K, et al. (2009) Relationship between blood pressure and outdoor temperature in a larger sample of eldery individuals: the three city study. Arch Intern Med 169: 78–80. doi: 10.1001/archinternmed.2008.512

56. Gliner JA, Raven PB, Horvath SM, Drinkwater BL, Sutton JC (1975) Man›s physiologic response to long-term work during thermal and pollutant stress. J Appl Physiol 39: 628–632.

57. Qian Z, He Q, Lin HM, Kong L, Bentley CM, et al. (2008) High temperatures enhanced acute mortality effects of ambient particle pollution in the "oven" city of Wuhan, China. Environ Health Perspect 116: 1172–1178. doi: 10.1289/ehp.10847

58. Bankir L, Perucca J, Weinberg MH (2007) Ethnic differences in urine concentration: possible relationship to blood pressure. Clin J Am Soc Nephrol 2: 304–312. doi: 10.2215/cjn.03401006

59. Donde A, Wong H, Frelinger J, Power K, Balmes JR, et al. (2012) Effects of exercise on systemic inflammatory, coagulatory, and cardiac autonomic parameters in an inhalational exposure study. J Occup Environ Med 54: 466–470. doi: 10.1097/jom.0b013e318246f1d4

60. Haskell WL, Lee IM, Pate RR, Powell KE, Blair SN, et al. (2007) Physical activity and public health: updated recommendation for adults from the American College of Sports Medicine and the American Heart Association. Circulation 116: 1081–1093. doi: 10.1161/circulationaha.107.185649

Citations

CHAPTER 1

Saleh Zakaria, Nadhir Al-Ansari, Yaseen T. Mustafa, M. D. J. Alshibli, Sven Knutsson, Macro Rain Water Harvesting Network to Estimate Annual Runoff at Koysinjaq (Koya) District, Kurdistan Region of Iraq, http://dx.doi.org/10.4236/eng.2013.512117.

CHAPTER 2

Amin MT, Nawaz M, Amin MN, Han M (2014) Solar Disinfection of Pseudomonas aeruginosa in Harvested Rainwater: A Step towards Potability of Rainwater. PLoS ONE 9(3): e90743. doi:10.1371/journal.pone.0090743

CHAPTER 3

Kaushik R, Balasubramanian R, Dunstan H (2014) Microbial Quality and Phylogenetic Diversity of Fresh Rainwater and Tropical Freshwater Reservoir. PLoS ONE 9(6): e100737. doi:10.1371/journal.pone.0100737

CHAPTER 4

Ouyang W, Chen S, Cai G, Hao F (2014) Dryland Soil Hydrological Processes and Their Impacts on the Nitrogen Balance in a Soil-Maize System of a Freeze-Thawing Agricultural Area. PLoS ONE 9(7): e101282. doi:10.1371/journal.pone.0101282

CHAPTER 5

McDonald RI, Girvetz EH (2013) Two Challenges for U.S. Irrigation Due to Climate Change: Increasing Irrigated Area in Wet States and Increasing Irrigation Rates in Dry States. PLoS ONE 8(6): e65589. doi:10.1371/journal.pone.0065589

CHAPTER 6

Jones BE, Krupa M, Tate KW (2013) Aquatic Ecosystem Response to Timber Harvesting for the Purpose of Restoring Aspen. PLoS ONE 8(12): e84561. doi:10.1371/journal.pone.0084561

CHAPTER 7

Chase Grey JN, Kent VT, Hill RA (2013) Evidence of a High Density Population of Harvested Leopards in a Montane Environment. PLoS ONE 8(12): e82832. doi:10.1371/journal.pone.0082832

CHAPTER 8

Anil Kumar Misra, Climate Change and Challenges of Water and Food Security, doi:10.1016/j.ijsbe.2014.04.006

CHAPTER 9

Barbosa CMG, Terra-Filho M, de Albuquerque ALP, Di Giorgi D, Grupi C, et al. (2012) Burnt Sugarcane Harvesting – Cardiovascular Effects on a Group of Healthy Workers, Brazil. PLoS ONE 7(9): e46142. doi:10.1371/journal.pone.0046142

Index

A

Agricultural area 80, 81, 82, 87, 112, 114, 119, 123, 124, 125, 128
Agricultural development 120
Agricultural practice 81
Agricultural production 4, 110, 226, 228
Agriculture 198
Air pollution 258, 274, 275, 276, 277, 278, 279, 280, 281, 282, 283
Air temperature 226
Antimicrobial activity 27
Aquatic system 52, 53, 68
Arabic Maghreb Union (AMU) 230

B

best management practices (BMPs) 139
Body mass index (BMI) 263

C

Canopy cover 144, 171
Chemical analyse 56
Commercial timber 137, 139, 140, 151, 183
Cultivation 198
Curve number (CN) 5

D

Digital Elevation Model (DEM) 5
Disinfection system 26, 32, 41, 44, 46

Dryland 79, 80, 81, 82, 83, 84, 85, 86, 87, 88, 91, 92, 98, 99, 100, 101, 102

E

Ethanol mill 259
Evapo-transpiration (ET) 80, 91, 92, 97

F

Fieldwork 199
Flat crop circle 203
Food security 225, 226, 229, 246, 247, 248
Forearm blood flow (FBF) 261
Freshwater 51, 52, 53, 54, 65, 66, 68, 69, 72, 75, 76
Fresh water resource 225, 226, 228

G

General circulation model (GCM) 112
General circulation models (GCMs) 111
Generalized estimating equation (GEE) 263
Global Climate Models (GCMs) 226
Greenhouse 110, 115, 116, 122, 129
greenhouse gases (GHG) 227
Ground cover 145
Ground water 3, 7
Groundwater 226, 228, 230, 231, 234, 235, 237, 238, 240, 241, 242, 245, 249, 250, 252, 253, 254, 255

H

Harvested rainwater 26, 27, 28, 31, 44, 45
Harvest period 262, 263, 264, 265, 266, 267, 268, 270, 272, 273, 274, 275, 276
Harvest season 259
Health problem 237
Heterotrophic plate counts (HPC) 52, 53
High humidity 55
Human health 52, 72
Hunting cover 198, 203
Hunting pressure 214
Hydrological condition 236, 250, 254
Hydrological process 80, 81, 82, 84, 99, 101

I

Indian Space Research Organization (ISRO) 227
Interquartile range (IQR) 263

J

Judicious water 3

L

Land farm 233
Leaf area index (LAI) 85
Linear mixed effects (LME) 150

M

Markov-Chain Monte Carlo (MCMC) 203
Microbial quality 52, 59
Middle East and North Africa

(MENA) 2
Minimal risk 210
Morphology 142, 191
Muscle sympathetic nerve activity (MSNA) 261

N

National park 209
Natural precipitation 229
nitrogen (N) 79, 80

P

Particulate matter (PM) 52
Particulate Matter (PM) 262
Plant organ 81
Policymaker 110
Polluted water 26
Population number 198, 209, 210, 212
Population source 212
Potential evapotranspiration (PET) 113
Predominant method 258
Primary production 151, 169

R

Rainfall 2, 3, 4, 5, 6, 8, 9, 10, 11, 15, 18
Rainwater 2, 3, 4, 5, 10, 11, 12, 51, 52, 53, 54, 55, 56, 57, 58, 59, 60, 61, 62, 63, 64, 65, 66, 67, 68, 69, 70, 71, 74, 75
Resident population 205
Root depth (RD) 85
Rubber tire 138, 144, 145

S

Sampling area 202
Seasonal activity 258
Seoul National University (SNU) 29
Small dam 4
Soil Aquifer Treatment (SAT) 226, 250
Soil Conservation Service-curve number (SCS-CN) 2, 5, 11
Soil erosion 4
Soil hydrology 81, 82, 101
Soil moisture 144, 149
Soil water storage (SWS) 89, 90
Soil water (SW) 80, 81
Solar collector 27, 28, 35, 37, 44, 45
Solar radiation 138, 139, 140, 144, 147, 148, 150, 151, 169, 170, 171, 174, 175, 183
Spatially explicit capture-recapture (SECR) 198, 201
Spot pattern 202
Standard deviation (SD) 263
Standard error 155, 158, 161, 164, 166, 168, 170, 180, 181, 182
Statistical model 109, 111, 115, 129, 130
Stream channel 147, 148
Stream temperature 139, 140, 150, 151, 154, 159, 161, 165, 171, 172, 174, 175, 176, 178, 179, 187, 193
Study area 200, 201, 211
Sugarcane 258, 259, 262, 263, 264, 266, 272, 273, 274, 275, 277, 279, 282
Sunlight 26, 27, 28, 29, 31, 32, 33, 35, 37, 39, 44, 48

T

Tillage process 83
Topography 202, 211
Total nitrogen (TN) 83
Treatment area 144, 145, 149, 150, 155, 158, 161, 164, 180, 181

U

United States Department of Agriculture (USDA) 112
United States Geological Survey (USGS) 112

V

Velocity meter 146

W

Water policy 111
Water resource 226, 228, 229, 231, 233, 234, 235, 236, 237, 244, 249, 250, 251, 253, 254, 255
Water scarcity 2, 3, 4
Watershed modeling system (WMS) 2, 5
Water source 27, 46, 80, 81, 83, 87, 91, 98, 101
Water temperature 26, 27, 33
Wet Bulb Globe Temperature (WBGT) 262